RECYCLING INDIAN CLOTHING

RECYCLING INDIAN CLOTHING

Global Contexts of Reuse and Value

————————LUCY NORRIS

INDIANA UNIVERSITY PRESS

Bloomington & Indianapolis

This book is a publication of

Indiana University Press
601 North Morton Street
Bloomington, Indiana 47404-3797 USA

www.iupress.indiana.edu

Telephone orders   800-842-6796
Fax orders   812-855-7931
Orders by e-mail   iuporder@indiana.edu

⊗ The paper used in this publication meets the minimum requirements of
the American National Standard for Information Sciences—Permanence
of Paper for Printed Library Materials, ANSI Z39.48-1992.

Manufactured in the United States of America

Library of Congress Cataloging-in-Publication Data

Norris, Lucy, [date]
  Recycling Indian clothing : global contexts of reuse and value / Lucy Norris.
     p. cm. — (Tracking globalization)
  Includes bibliographical references and index.
  ISBN 978-0-253-35501-0 (cl) — ISBN 978-0-253-22208-4 (pb) 1. Clothing
and dress—India. 2. Clothing and dress—India—Environmental aspects. 3.
Clothing and dress—India—Remaking. 4. Recycled products—India. 5. Used
clothing industry—India. 6. Culture and globalization—India. I. Title.
  GT1460.N67 2010
  363.72'82—dc22

                                                        2009053205

1   2   3   4   5   15   14   13   12   11   10

For FLORIAN

CONTENTS

ACKNOWLEDGMENTS

In India, fieldwork could not have been conducted without the friendship and patient cooperation of the residents of the Progressive Housing Society and the Waghri dealers in Raghubir Nagar, who all remain anonymous. Lata and Suman Sharma and Rimly Bezbaruah welcomed me into their family lives. Bharati Chaturvedi of the NGO Chintan generously suggested contacts and new perspectives when I arrived, while Sunita Bhaduria, Shiv Kumar Malhotra, and George Samtabhai provided invaluable research assistance. Jasleen Dharmija, Santosh Desai, Achin Ganguly, Famida Hanfee, Anupam Jain, Jyotindra Jain, Madhu Jain, Rohini Kosla, Ajit Kumar, Anamika Pathak, Amba Sanyal, Arun Shah, Kishore Singh, Shobha Deepak Singh, Pooja Sood, and Laila Tyabji all generously gave their time to discuss such topics as Indian fabrics, development, recycling, consumption, and the fashion business. Sonam Dubal, Mrs. Patel, and Simon Wilson in particular shared their experiences as designers of fashions made from recycled clothing.

Early research was made financially possible through a Research Studentship from the UK Economic and Social Research Council (R00429834592), the Royal Anthropological Institute's Firth Award for 2002, and two grants from the University of London Central Research Fund in 2000 and 2004 to carry out further fieldwork in India. Subsequent writing was supported by an ESRC Postdoctoral Research Fellowship (PTA-026-27-0013) and the ESRC *Waste of the World* project (RES 000-23-0007). Documentary photography and additional research were funded by a British Academy Small Research Grant 2004 (SG 38685).

Susanne Küchler has been an ever-helpful mentor who patiently helped to make sense of the notes and chapters as this study took shape, and whose work on the material mind continues to inspire. Mukulika Banerjee generously gave advice on setting up the initial research project in Delhi, provided invaluable contacts and assistance in the preparation of fieldwork plans, and offered insightful comments on subsequent drafts.

Jonathan Parry and Deborah Swallow provided useful comments on an earlier version of the manuscript. Robert Casties, Haidy Geismar, Rob Irving, Patrick Laviolette, Frances Lloyd-Baynes, Jean-Sebastien Marcoux, Danny Miller, Kaori O'Connor, Claire Sussums, Graeme Were, and Diana Young offered constructive criticism of draft chapters, and I benefitted from the support offered by members of the Department of Anthropology at University College London.

I would like to thank Bob Foster for including the book in the Tracking Globalization series, and Rebecca Tolen for editorial support at Indiana University Press. Both have been extremely patient, and generous with suggestions for improvements. An insightful report from an anonymous reviewer led to a thorough revision of the first half of the manuscript, and I am grateful to that reviewer for the opportunity to improve the contextual information and the organization of the text. Shoshanna Green made numerous helpful suggestions during copyediting; any remaining errors and inconsistencies remain my own.

From a previous career working at the Horniman Museum, I was moved by Keith Nicklin's passion for ethnography to do post-graduate study of anthropology at University College London. Later, Tim Mitchell accompanied me on a trip to Delhi in 2004 to take a series of photographs that have opened up new ways of thinking about my work, as well as making it accessible to a wider audience, and I am grateful to the Horniman Museum for hosting a photographic exhibition of his work, *India Recycled*, during 2008–2009. My parents, family, and friends have been immensely supportive, and my husband Dirk has spent many a long day in second-hand markets and recycling factories; it is his joyful enthusiasm for visiting India both before and after the arrival of our young son, Florian, which has made this project so much fun.

RECYCLING INDIAN CLOTHING

FIGURE 1. In Raghubir Nagar, a suburb of Delhi where thousands of people earn a living recycling used clothing, a new temple to the god Ramdev is being built. Local dealers in cast-offs regularly contribute a small percentage of their profits, thus the temple is both a symbol of transcendence and a material manifestation of the value to be extracted from old cloth. PHOTO COURTESY OF TIM MITCHELL

1 RECYCLING INDIAN CLOTHING
The Global Context

Values in Indian Clothes

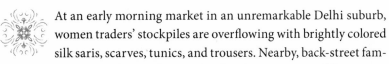 At an early morning market in an unremarkable Delhi suburb, women traders' stockpiles are overflowing with brightly colored silk saris, scarves, tunics, and trousers. Nearby, back-street family workshops are producing thousands of cushion covers, wall hangings, halter-neck tops, dresses, and skirts embellished with woven, printed, tie-dyed, and embroidered designs in Indian sari fabrics, all destined for the export and tourist markets. These are the subjects of this research on trans-national material flows, but the story goes beyond the production of fashion and furnishings in north India and its relationship to global consumption. This book tells a much more surprising and little-known tale, that of changing indigenous practices of disposal, reuse, and recycling of local clothing and the burgeoning market in used textiles in a rapidly changing Indian consumer society.

Starting with the wardrobes, trunks, and suitcases of middle-class metropolitan households, bulging with old and unwanted clothes, the study looks at why people in India need to get rid of clothing, and how the way in which they do this is changing as new social worlds are developing. Images of unwanted surplus and the problems of what to do with waste clothing are becoming familiar to consumers in the West, but related recycling systems in newly developing economies have as yet attracted little attention. Cloth and clothing is never just thrown out as rubbish in India. It is too replete with social meaning to be wasted until it is literally falling apart. Used clothing is still a valuable resource; it can always be strategically gifted, used up, or exchanged for something more desirable. Unlike the plastics, metals, and glass recovered by ragpickers, textiles are rarely found among the garbage dumps for domestic waste; unwanted clothing follows a different route out of the house altogether.

Treasured pieces can be preserved for favorite younger relatives, suitable, serviceable clothes gifted to a maid, and rags reused in the house. Used clothing is made to work, to produce value for the home. But since the liberalization of India's economy in the early 1990s and the rising consumption of clothing and fashion, what happens to the increasing surplus of clothing that is "too good for the maid"? Women wonder what to do with the growing piles of good-quality clothing rendered unwearable by the vicissitudes of daily life, which represent too great an investment to be simply given to a servant and are too valuable to waste by leaving them sitting unused in cupboards at home. The most problematic category of all is that of old silk saris, at once the most valuable clothing in the home and potentially the most redundant.

Domestic and familial recycling of clothing, understood as a repetitive practice that can conserve both economic resources and sentimental value, is a way to extend love or protection to relatives and servants, and can also be skillfully used to negotiate status and value within these relationships. Old clothing can also be recommoditized, usually by bartering surplus clothing for pots, plates, and kitchen utensils. Such bartering is a theatrical exchange on the threshold of the home, and the moment when the most intimate of personal possessions are pared from the body and cast out of the wardrobe.

Hidden out of sight in warehouses, factories, workshops, and the backstreets of slum neighborhoods, vast quantities of old, unwanted clothes that have been bartered for pots are recycled for the local and global markets. Alongside the well-known centers of Indian textile production that supply the international textile market are more covert recycling industries, which convert discarded remnants of old clothing into attractive products for both the home market and further afield.

This transformational creation of a second life for old garments itself constitutes an important trans-national flow of materials, an efficacious waste stream of fiber, cloth, color, and pattern that bears endless potential to re-create value, yet one that remains largely within an unacknowledged global underworld associated with dirt and decay. The manipulation of complex concepts of materiality and value in a rapidly changing world lies at the heart of the cultural transformation of waste; old clothes are multi-layered, composite objects made up of fibers and fabrics, shapes, colors, and patterns, overlaid with memories, smells, and associations. A singularized garment whose history recalls for one woman a lifetime of family life (Kopytoff 1986) may reenter the market stripped bare, leaving only color and fabric to be reinvented once again as an ethnic textile cushioning a Western woman's sofa.

This story of old cloth delves deep into the highly localized contexts of the international trade in cast-off Indian clothes, revealing the series of processes through which clothing is rejected from the Indian wardrobe, commodified in Delhi markets, and transformed through creative imagination into fashion and furnishings that appear in domestic markets for urban consumers, in the regional hinterland, and in the international clothing trade. From individual examples of recycling in the home to major market transactions involving thousands of silk saris, the study follows ever-expanding circuits of production and exchange, tracking commodities as they move through multiple sites along trans-national chains.

Although it begins as a study of indigenous clothing recycling practices, my research also extends systematically back to the West through the international market for recycled saris. I follow the hidden flow of used clothing from the developing world to the West, tracking the trade through production, consumption, disposal, and transformation. I draw initially on Appadurai's notion of the social life of things (Appadurai 1986b), but I go on to consider the disassembly and reassembly of objects, viewing pieces of cloth in all forms as agents in transformatory processes, examining the cultural slippages whereby a piece of Indian textile translates into a Western fashion item.

This study investigates the meaning of clothes as material objects in India today, and poses the question of what happens to garments that people no longer need or want. Turning that question around, it also addresses the potentiality of material things in India and asks what it is that a discarded piece of cloth can become, and why. Examining why some people discard and how others transform will show us that such practices and their associated value systems depend upon fundamental perceptions both of the materiality of people and things and of their mutual constitution. Unraveling and remaking material things such as textiles engenders the breaking and remaking of the social relations with which they are associated, making recycling a socially transformative practice.

We need an understanding of cultural attitudes toward surplus and the potential of waste for creating value if we want to understand the subtler dynamics of such transformations. Korom's essay on recyclia focuses upon the resourcefulness of transformers of waste, and provides an account of the social difficulties faced by poor upper-caste recyclers who are polluted through their work of transforming rubbish, such as tin cans, into recyclia (Korom, 1998). The study of the recycling of cloth brings into focus Indian beliefs concerning the nature of identity, personhood, and sociality, ideas about possessions, gifting and exchange, and the influence of cosmological beliefs. Indian society is still partially a cloth economy, where cloth is both a currency and a means of incorporation (see Stallybrass 1993), which has profound implications for disposal practices. As cloth is gifted and exchanged throughout life-cycle rituals, in religious and personal celebrations, it creates social relationships that have to be continuously reinforced as the cloth wears thin; the processes involved in reuse and recycling are here understood as the simultaneous transformation of trash into treasure and the remaking of social ties that constantly threaten to unravel or tear.

The transformation of clothing, through its material properties of decay and its potential for renewal through cutting and stitching, enables the transformation of the person and social relations. Gell demonstrates that the object has no intrinsic nature, independent of the relational context in which it is embedded; Gell treats objects as a system of action, intended to change the world rather than encode symbolic propositions about it. Thus his anthropological theory of objects "merges seamlessly with the social anthropology of persons and their bodies" (Gell 1998, 6). The logical outcome of a theoretical approach which focuses upon the efficacy of materials as relational constructs (for example, Strathern 1988, 1999) points to the role of the dispersal, loss, and destruction of objects in processes of memory formation, forgetting, and remaking (Küchler 1992, 2002).

Deeply held beliefs regarding cloth were revealed during the political struggle for independence, and they were utilized by Gandhi in his Swadeshi (Home Industry) campaign (Bayly 1986; Cohn 1989; Bean 1989). Bayly analyzes the exchange of cloth in the Mughal era to extract the underlying rationale concerning cloth transactions, the manner by which cloth not only fixed and symbolized social and political statuses and changes in them but was also able to transmit holiness, purity, or pollution to the wearer. Bayly uses the term "bio-moral substance" to convey the Hindu sense of the constant substantivist mingling of the essences of persons, substances, morals, and actions.

As they do today, cloth transactions and the wearing of new clothes marked every major life-cycle ritual, every religious festival, and every affirmation of political alliance in the Mughal period. Bayly identifies three uses of cloth in the social process: for symbolizing status or recording change in status; its magical or transformative use, "in which the moral and physical being of the wearer/recipient was perceived to be actually changed by the innate qualities of the cloth or the spirit of substance it conveyed"; and its use as a pledge of future protection (Bayly 1986, 286). During the development of proto-capitalism in the colonial period, although the meaning and functions of these exchanges changed, Bayly concludes that cloth was only partially commoditized. It continued to retain the "spirit of the gift," the quality of the producers and traders who came into contact with it.

In contemporary north India, similar beliefs about the capacity of cloth to transmit the essence of people, places, and times continue to be held, and the way in which cloth is gifted, kept, and handed on contributes to the ongoing development of personhood. Material flows in the form of consumption, use, and, critically for this study, *discarding* are crucial to the process of making and remaking social identities. The discussions about clothing in my own fieldwork echoed Lamb's findings in Bengal, where her informants

> viewed the sharing and exchanging of bodily and other substances—not only with other people but also with the places in which they live and the things they own and use—as vital to the ways they think about and define themselves and social relations. Parts of other people, places, and things become part of one's own body and person, just as parts of oneself enter into the bodies and thus the persons of others. (2000, 39)

The life cycle of cloth can be understood as a material flow through the household managed by women; it is vitally implicated in fundamental concepts of purity and pollution, auspiciousness and inauspiciousness, rubbish and value. Cloth may never be thrown out as rubbish, but it certainly becomes surplus. Unwanted, it needs to be got rid of from time to time, both to maintain a state of domestic and personal cleanliness and present a good image to the world, and to realize its maximum potential value. Old clothes must be evaluated for their usefulness as clothing in creating an image, as sentimental reminders of another person or occasion, as potential gifts for a younger relative or servant, as resources for recycling, as rags around the house, or as items to be exchanged for things of greater value. And who one hands them on to, and in what context, can have as much influence on their subsequent value and meaning as where they have come from. Korom's work on Indian recyclia highlights some of the cultural problems of dealing with waste, in which concern for the environment is not a factor (Korom 1996). In India, materials are usually reused because of economic necessity and practical utility, and doing so involves scavengers and makers in work that is polluting because it entails contact with refuse. This pollution affects the status of both the recyclers and their products.

The recycling of cast-off clothing in India continuously creates social persons in relation to each other. In a culture where used clothing, which has been worn on the body, continues to be perceived by many upper-caste Hindus as a potential source of pollution, the giving and receiving of old garments creates hierarchical distinctions that divide the higher giver from the lower receiver. The giver needs to give in order to establish his or her superiority, and this is arguably one of the most important uses of old clothing. Similarly, old clothing may have certain negative associations and be considered inauspicious. But once the clothes are commodified, recyclers as cultural brokers work to remove traces of biographical information concerning their former lives from them in order to maximize their value to those outside the indigenous belief system. Foreign buyers are instead tempted by images of opulent styles and luxurious fabrics, small pieces of India to take away and incorporate into their homes. Traders across north India have opened up a huge market for waste cloth abroad where there was none before, taking advantage of differing cultural attitudes to waste and leftovers to move goods into alternative value regimes, yet simultaneously re-creating such differences anew in the process. Links are not made between buyers and sellers: consumers in the West generally do not know where their Indian sari products have come from, and middle-class Indian women usually do not realize where their old saris go to. These businesses in the informal economy operate largely below the level of state regulation, seizing opportunity as soon as it comes.

Overconsumption and Riddance in Delhi

The contemporary problem of overconsumption, excessive wardrobes, or surplus clothing in non-Western countries has received scant attention from anthropologists. Those working in non-Western cultures have more usually described the value of keeping "traditional" textiles, their importance for social identity and exchange, and either their preservation though a person's life or at the very least their ritual destruction (e.g., Weiner and Schneider 1989). Contemporary accounts of clothing across the developing world focus upon design, fashion, and fusion, consumption patterns and market economies, but not yet upon the impact of increasing consumption on former cultural strategies of getting rid of unwanted clothes at the end of their useful lives. Piles of unfashionable, unsuitable clothing, not yet worn out but no longer wearable, are a newer phenomenon as emergent economies such as India's "liberalize," "open up," and, above all, grow.

There are no available statistics on the amount of used clothing cast out from Indian homes, and it is not yet possible to determine to what extent the numbers of unwanted garments are increasing nor the overall size of the informal economy that processes them. But, as this research will show, it is reasonable to assume that old clothing follows the trends of other material waste, which is increasing in line with burgeoning consumption. The research reported here began in 1999, just eight years after the Indian economy was deregulated in 1991, and at the beginning of the contemporary consumption boom.[1] Such a huge expansion in the economy has led to the well-known emergence of a large urban middle class with a significant level of income, both declared and undeclared. Béteille includes in the term "urban middle classes" the "intelligentsia, professionals and service classes" (Béteille 1997, 151). In 1997 he estimated that the upper stratum of civil servants, managers, and higher professionals would number hundreds of thousands, and if one were to include schoolteachers, clerks, and other white-collar workers, it would expand to 50 to 75 million.

A year later, Varma estimated their numbers to be at least 200 million in a total population of one billion (1998), a figure which tallies with those quoted by the marketing managers of international advertising agencies at the time,[2] reflecting the perceived market for consumer products by the end of the century.[3]

As part of an ongoing restructuring process, middle-class Indian families are moving from extended families to nuclear ones and relocating to mega-cities; they have more disposable income, give and receive more clothes, purchase more fashionable items, and have fewer avenues for disposal of them. Old clothing used to be handed on to younger, poorer relatives or servants. (Crucial to this handing on of clothing is the fact that saris—and *dhotis,* men's loincloths—are untailored and can be wrapped around any figure, and that even the women's outfits called *salwar kamiz* or "Punjabi" suits are usually made with wide seams and can be taken in or let out as necessary.) Women's wardrobes overflowing with clothing are themselves a materialization of the increased connectivity between India and the global economy and its concomitant effects, often with more clothing coming into the home than can be properly dealt with by former strategies of domestic reuse and recycling. Often clothing is not fully worn out before it must be got rid of, to make space for new images and more fashionable forms generated by the increasingly rapid changes in fashion.[4] In addition, the increasing adoption of Western clothing is in itself leading to more redundancy, as clothing is tailored to certain body shapes and sizes, and increasingly differentiated by fashion trends and styles appropriate for age groups. The styles of clothing owned, their provenance, and their increasing number attest to the growth in prosperity of givers and receivers of cloth, and to networks of production and consumption extending well beyond India.

As it moves through pathways of use and exchange, reuse and recycling, clothing is transformed into cloth and back again.[5] If cloth is perceived as a "second skin" in the formation of an embodied self, then the shedding of skin is the moment where people are most able to re-create and reveal themselves anew. A couplet from the *Bhagavad Gita* was often quoted to me by friends and informants. Krishna says to Arjuna, "Just as a man giving up old worn out garments accepts other new apparel, in the same way the embodied soul giving up old and worn out bodies verily accepts new bodies" (*Bhagavad Gita* 2:22).

Exporting Recycled Indian Clothing

The domestic disposal of clothing highlights the importance of cultural perceptions of materiality to understanding exchanges of value at every level of trade. The findings at the local level can be applied equally well to national and international settings, through examining the usefulness of waste materials in the creation of social relationships at the level of the individual, the middleman, and the global trader. The potential of sacrificed clothing to become a new piece of cloth, a textile commodity to be traded in the global market, reveals the potential of the destruction and remaking of material things to serve as a technology of making and remaking self and other in a fluid, ever-changing global landscape that transcends cultural borders.

Having begun by considering the local and domestic context, this book then shifts its focus to the marketplace and the activities of traders and middlemen who overturn existing regimes of value. It charts the entrepreneurial technologies of transformation, as clothing is washed for resale to the poor, ripped up for use in the international machine wipers market, or cut up and restitched for different cultural markets across the continent. Decorative silk saris are the most lucrative material form in the market, and they may be used to create new products for foreign markets, with middle men often working alongside foreign travelers and informal marketers.

During the early 2000s, not only was mainstream fashion in the West heavily inspired by Asian design elements, but shops and stalls across the UK were full of "sari cushions," "sari skirts," and halter-neck sundresses made out of Indian silk sari fabrics quite different from the block-print cottons, ethnic mirror-work embroidery, and cheesecloth of the previous fashion cycles. In 2008, recycled saris were being sold on eBay for £1 each, and women's recycled sari bags were marketed online by businesses such as Greener Style.[6] Some of those involved in recycling saris into UK designer fashion are entrepreneurs working under the umbrellas of ethical fashion, social accountability, and sustainability, such as Sari UK, which recycles saris from the Indian community in the UK, and Jeannette Farrier, one of many designers who use old saris from India itself.

In Germany, an importer with a chain of shops called Chapati Design sells ethical sari clothing made in Delhi and displays recycled saris alongside new pieces in his shops. Small firms such as these, often promoting Fair Trade practices and informed consumer choices, often outline the benefits that buying their products will have on the makers of such clothing, who may be located in workshops in, for example, north India. These companies use their websites and promotional material to show the conditions the clothing is made under. But other companies selling mainstream women's clothing that includes recycled silk saris appear to utilize saris simply as a fashionable resource without any ethical added value; one called Namaste UK offers

> Recycled Silk Sari Skirts, Dresses and Trousers: Assorted colours of beautiful recycled sari made into floaty clothing—fit for parties, summer holidays in exotic places, and just lazing about looking fantastic![7]

The origins of most of these recycled saris and the means through which they have been obtained remain elusive; this information is rarely detailed,[8] and this reticence suits both sides of the transaction, for the profits to be generated from recycling very cheap source material and selling it on as Western fashion are also not revealed, a typical characteristic of many recycling industries. Most recently, Indian dealers have started to market old saris as "vintage textiles" online on trading sites such as eBay, thus directly connecting the larger players in the market with individual foreign buyers and retailers and eliminating the foreign importer completely. Selling them as one-offs, they open up a whole new market for customers who collect saris or reuse the fabrics for their own creative projects.

Global Textile "Waste" and Used-Clothing Markets

In the West, newspaper reports, TV programs, and popular media daily point at the amount of "waste" produced around the globe and its negative consequences. The generation of material classified as waste increasingly poses a problem for those civic bodies responsible for its elimination, those concerned about the effects of various means of disposal upon the environment, and those trying to improve the health and well-being of those who suffer from those effects. We are also aware that refuse from far-flung sites of production and consumption is traded across world markets, and that some waste materials supply legitimate reprocessing plants, while other impoverished communities are dependent upon waste that is illegally dumped.

Textile waste has been growing exponentially in the West in the last decade. In the United Kingdom, approximately 1.9 million metric tons of textiles and footwear were consumed in 2003; 1.2 million metric tons of used clothing and soft furnishings were disposed of as waste, and a further 303,000 metric tons of textiles were reprocessed by the secondary textile industry for reuse or recycling (Oakdene Hollins Ltd. et al. 2006). In its *Waste Strategy 2007,* the UK government has identified textile waste as the fastest growing domestic wastestream after aluminum (DEFRA 2007, annex D, 141). An increasing textile wastestream testifies to the UK's failure to push waste higher up the five levels of the waste disposal hierarchy (with prevention at the top, followed by reduction, reuse, recycling, and disposal at the bottom) as more and more cast-off clothing in the UK falls to the lowest level of rubbish (DEFRA 2007).

This emergent problem is predominantly understood and portrayed as the West consuming too much fashion, and previous strategies for dealing with old clothes being stretched to breaking point (DEFRA 2007). The problem is exacerbated by the reduced lifespan of many newly manufactured goods and by ever-shorter fashion cycles. Poorer-quality, cheaper clothing is being imported into Europe and the U.S. in ever-increasing volumes as fashion cycles shorten to barely four to six weeks, and, according to organized charities and textile recycling merchants, it is becoming harder to find economically viable ways of recycling low-grade textiles rather than sending them to a landfill (Oakdene Hollins Ltd. et al. 2006).

The pressing questions in the West are now perceived to be how to reduce textile disposal through encouraging extended use in the first place, how to reuse clothing through redesign or simply resale, and how to increase recycling through disassembling component parts of a garment and bringing them back together again in a new form. What values can be called upon to maximize a textile's potential after it has ceased to be worn as clothing? This is a conspicuously difficult problem in the West simply because clothing is so closely associated with fashion. The very term "fashion" implies that clothing is redundant when it is no longer stylish rather than when it is no longer usable.

Traditional solid waste management specialists undertook "cradle to grave" life-cycle analyses—waste was the end of the line. But in recent years we have seen the introduction of an alternative, entrepreneurial conception epitomized by McDonough and Braungart's 2002 book *Cradle to Cradle*, which envisions materials as resources to be assembled and reassembled into new products as required. Their work appeared at the same time as Murray's *Zero Waste*, whose back cover informed readers, "As a pollutant, waste demands controls. As an embodiment of accumulated energy and materials it invites an alternative." In the United Kingdom, waste management is fast being reconceptualized as "resource recovery" in various commercial and civil sectors (Lisney, Riley, and Banks 2003–2004; Riley et al. 2005).

Viewing waste as a set of potential values waiting to be recombined and reactivated in new forms resonates with classic anthropological work on gifting and exchange, in particular work on value transformation and materiality (e.g., Munn 1986). The cultural contexts in which such transformations take place become highly complex when waste material crosses international borders. Research on the spaces of recycling adds to the critique of the simple dichotomy between the gift, characteristic of noncapitalist societies, and the commodity proposed by Mauss (1954) and subsequently developed by Gregory (1982) and Carrier (1995). The dichotomy is clearly inappropriate for the more marginal contexts in which secondhand commodities are sold, such as charity shops (Gregson, Brooks, and Crewe 2000), car boot sales (Gregson and Crewe 1997), and garage sales (Hermann 1997). For the latter, Hermann notes that "the metaphor of the market—that is, all the apparent trappings of buying and selling—simultaneously cloaks and facilitates a web of transactions that are often as much as or more socially engaged as economically rationalized" (Hermann 1997, 912).

The study of clothing recycling practices highlights, above all, the shifting concepts of value as objects move in and out of a variety of acknowledged flows of goods, muddying the waters of supposedly clear distinctions between things steeped in personal meaning and impenetrable goods advertised in the market. When cloth moves in and out of the market in India, the "spirit of the gift" threatens to adhere to its fibers, and this problem will be returned to in the discussion of barter in chapter 5. A study of charity shop workers in the UK revealed that clothing was valued more highly by middle-class consumers if specific bodily traces of individual former owners were removed as much as possible, but the clothing still looked old and reused rather than new (Gregson, Brooks, and Crewe 2000). Whereas one might expect charity shop workers to decontextualize the clothing, in fact the clothes require a lateral shift, a recontextualizing; the fact that they were "genuinely" old, and therefore possibly marketable as "vintage," was the new source of their worth. In contrast, in the sale of used baby clothing, knowledge of the former middle-class owners and the ability to reaffirm the clothes' relational context are likely to increase its value if they are balanced against a proper degree of separation (Clarke 2000). The thread running through all these recent studies is that work is required to make objects suitable for further exchange, once they are no longer suited to their initial use. It is this work that creates new value for used clothing.

Clothing design, manufacture, and consumption occur within complex global economic matrices, where it can be difficult for the end consumer to ascertain the multiple origins of the constituent elements of garments (Rivoli 2005). We simply don't know where things come from. Non-European countries enter the supply chain economics of recycling and reuse largely as cheaper sites of production. In particular, the relationship of textiles and the non-Western world is usually perceived as one where mass-produced clothing is imported into the West from newly developed countries such as China, and eagerly consumed before being thrown out as low-value waste (a.k.a. the "Primark factor" or "Wal-Mart factor").

This relationship also exists in many examples of "ethical fashion" produced in developing countries. The makers of Worn Again training shoes, for instance, have their UK-styled footwear manufactured in China from recycled Western material as part of a partnership with ethical fashion brand Terra Plano, although major efforts are now being made to bring all production back to Europe. From Somewhere, a highly fashionable company that recycles off-cuts from chic Italian knitwear companies, has manufactured its *haute couture* garments in the marginal European economy of Poland, where labor has remained relatively cheap; and so on. Some designers make the recycled origins of their garments explicit through the obvious restyling of component materials, such as the reuse of disassembled men's suits with identifiable sleeves, cuffs, and lapels by Junky Styling, thus highlighting strategies of creative reuse as a fashion statement in themselves. Others produce garments which look like familiar fashion items made from new materials but whose labels reveal the contextual information that the fabrics and fibers used are recycled.

In the last decade, research has been focused on the global trade in used clothing, through which Western cast-offs are reexported as second-hand clothing to poorer Third World economies in Africa, South America, and Asia. The trans-national trade in used clothing is not a new phenomenon, but it has reached an unprecedented scale.[9] Hawley (2006) has discussed the economics of the sorting procedures Western clothing is subjected to before niche export markets can be developed and exploited, describing the most sought-after antique clothing as "diamonds" to be dug up from the mass of waste cloth. Hansen (1994, 2000) has most thoroughly analyzed the life cycles of discarded clothing which has been exported to Zambia, where it is called *salaula* and highly regarded as a means of reconceptualizing fashion. Recycling clothing in India is both an indigenous practice and a new resource for global fashion, and its scale in India is enormous and largely unreported.

This book focuses on the reverse of the standard pattern, investigating the flow of clothing from a rapidly developing country into Western markets as well as domestic ones. This flow continues a series of domestic practices and extends them into the commercial sphere, providing a way for overburdened families to rid themselves of excess and supplying clothing to millions of India's poor. At the same time it engenders new opportunities, such as the reworking of material into fashionable items for the Western market through complex translations of value systems. In the process of detailing such practices, this study engages with important areas of anthropological enquiry such as the ontological relationship between personhood, material culture, and processes of attachment and riddance; concepts of value, dirt, and perceptions of materiality; the role of entrepreneurial translators in transforming clothing for local and international consumers; and their manipulation of value systems for end consumers.

The Structure of the Book

Chapter 2 takes a series of field research locations in Delhi and describes them in terms of the use of cloth, its exchange, and its agency in producing the character of those urban spaces. These contexts include the housing society where I resided in east Delhi, local weekly markets and Sunday markets, and wholesale traders' markets where used clothing is exchanged.

Chapter 3 addresses the central question of how clothing is valued by middle-class women in contemporary urban India. Taking examples from interviews and grounding them in extensive research among these women, the text explores the means by which clothing is acquired, used, stored, and organized. The focus is upon issues of consumption and the wearing of clothing, the importance of gifting cycles, and how these correlate with theories of the relational self ascribed to South Asians. The chapter explores views on the materiality of cloth and beliefs concerning its ability to transmit the essence of others, and asks how cloth is utilized strategically as a resource, both as clothing and as a component of a woman's wardrobe. The family wardrobe is conceived as a collection of "selves" that is managed by the woman in the household; the accumulation of clothing reflects the development of the social self. Clothing can be hoarded for sentimental reasons, and piled up to satisfy the requirement that a woman have an excessive amount; she becomes a "larger" person by acquiring more. Yet old clothes have become a burden: flats are getting smaller, nuclear families in cities are replacing extended households, and the fashion cycle is speeding up. Disposing of clothing is seen as necessary, as a skillful way of managing resources, and as an iconoclastic practice that sacrifices a part of the self and creates exchange value.

Chapters 4 and 5 address ways of getting rid of unwanted clothing. Chapter 4 focuses on the reuse of garments within the home and family, whereby the ties of love and affection and reaffirmed. Clothing is used as a protective device which reinforces the boundaries of the domestic sphere. Clothes are handed down to younger siblings and cousins, shared and swapped with sisters, cousins, and mothers, inherited from mothers and favorite aunts. When they can no longer be worn, they are cut up and made into children's clothing or baby wrappers, or are patched and quilted into furnishings for the domestic space. Old clothes can be considered welcome, even auspicious, resources; inauspicious cast-offs; or even polluting leftovers, *jutha,* depending on the relationship between giver and receiver. The offering of cloth and its acceptance or rejection is in fact constitutive of such relationships, and so approaches outside the family must be made with care. A common choice is to pass down clothing to family servants, simultaneous insiders and outsiders, who expect such gifts regularly. Such gifts will do for daily wear, but the dilemma is often faced of what to do with silk saris that are "too good for the maid."

Chapter 5 turns to the processes whereby clothes are commoditized by being bartered on the doorstep for kitchen pots, and features a case study of the problems faced by an elderly woman trying to get rid of a once precious waistcoat. It briefly looks at alternative commoditization strategies, including marketing clothes as antiques (which is, however, beyond the aspirations of most sellers) and burning them to extract gold and silver. The barter trade is ubiquitous, yet unacknowledged by most of the middle and upper middle classes. Women of the Waghri caste trade shiny new pots for bundles of old clothes, giving the clothing's owners "something for nothing" and understanding used clothes to be sacrifices. The chapter analyzes the choice of barter as an exchange mechanism and the role of the Waghri as outsiders who are able to remove rubbish from the household and the local community, thus ensuring that old clothing never returns unexpectedly. It discusses the appeal and symbolic value of pots in relation to the metaphors surrounding cloth, and examines the common pairing of pots and cloth as elements of a woman's dowry and of her domestic realm. The chapter concludes with a review of Indian conceptions of rubbish and value, and of opportunities to dispose of trash in return for treasure.

Chapter 6 describes the recycling of clothes that have been bartered to the Waghri women, and focuses upon the fate of silk saris as an example of trans-national trade. It analyzes how middlemen selectively buy up particular clothing types in the local market, where the traders take their daily goods. Options for its further use depend upon the perceptions of the material qualities of the cloth. Old cotton saris and *dhotis* are sold as rags on the international market; army uniforms become guards' clothing, and everyday wear is washed and sold on to the poor. Silk saris are the most lucrative find, as they can be transformed into fashionable furnishings and clothing and sold to tourists in India and abroad.

Successful transformations of saris into tourist goods play upon perceptions of Indian fabrics, with their silk material and decorative borders, as exotic. Their second-hand origins may be concealed, or the cloth may be advertised as "recycled." Western styles are skillfully copied by tailors and manufacturers, encouraged by the constant flow of visitors to well-known tourist sites across India. For some years, these garments have been sold in markets across Europe, the U.S., Australia, and Japan, and they have contributed to the rise in Asian influences on fashion that peaked at the millennium. The styles were picked up by leading fashion magazines and catwalk designers, and main-street retailers soon copied them. However, a very different strategy was adopted by young designers aiming for the elite market in India. Cultural prejudices against cast-off clothing have been challenged by sophisticated marketing; rather than concealing the clothing's origins, designers may deliberately feature old-fashioned motifs or use "distressed" silk, and labels stress both the environmental benefits of recycling and timeless "Indian" beliefs in reincarnation. These designers' clothes are now found on international catwalks.

The concluding chapter draws together the various themes developed throughout the book, and frames the practice of recycling clothing as a technology for remaking the self through iconoclastic destruction and exchange. Not just the self but ever-expanding units are remade in this way, as families, classes, communities, and states seek to rid themselves of unwanted images and remake themselves in new forms. The relationality embedded in cloth is constantly in danger of disintegrating, and requires renewal; cloth and its relatively short lifespan are therefore essential resources in the constant struggle to counteract long-term definitional practices rooted in the concrete. Thus examining the material properties of textiles reveals in fine detail how people and things really do create one another through exchange, allowing for temporary, shifting identities that can be continuously remade.

FIGURE 2. A dealer sells recycled jeans labels, old zips, cotton thread, and rolls of reject labels from the garment export trade at Raghubir Nagar wholesale market. PHOTO COURTESY OF TIM MITCHELL

2 FIELDWORK CONTEXTS

Clothing and the Domestic Sphere

The Progressive Housing Society

The research for this book was conducted whilst I was living in New Delhi for a year, from July 1999 to June 2000, and on subsequent visits in 2004–2005. During the year 1999–2000, I rented a flat in a cooperative housing society in a newly developing suburb on the east side of the River Yamuna, known as Trans-Yamuna. Amongst the dense building works, the remnants of villages remain along the banks of the river, where fruit and vegetables are still grown. Some older districts in Trans-Yamuna are distinctly upper-middle-class, with large private houses standing on single plots. But in the late 1970s the Union government decided to sell public land in order to encourage middle-class cooperatives to develop private housing projects that could offer affordable housing to supplement the inadequate public housing stock. By the millennium, land prices were soaring and developers fighting for high profits.

In the local market were hole-in-the-wall provisions shops and service providers, dairies, sweet makers and bakeries, and newer stores selling household linens, electrical goods, and home decorating supplies. There were also the local ready-made dress shops (including one that was owned by a neighbor of mine), shops selling export surplus clothing and cheap factory outlet remnants, internet cafés, and STD/ISO telephone booths, all catering to the middle-class owner-occupiers of the apartments and to those who serviced their lifestyle. There was constant development: during the year I would leave the area for a couple of weeks at a time and return to find the small shops bulldozed and larger stores being built on the same plot, roads being widened, slums cleared. Behind a façade of polished marble veneer, crumbling concrete shops and the houses of petty traders underwent a continuous cycle of destruction and rebuilding.

My flat was in a gated housing association like hundreds of others in the area. I will call it "The Progressive." Here I got to know my neighbors while learning Hindi,[1] and from here I traveled across the city, visiting wholesale markets, workshops in slums, and top-end boutiques in urban villages, all locations where my neighbors' cast-offs were recycled (as they were sometimes amazed to find out). In piecing together the economy of used cloth while working in the mega-city of Delhi, I followed the extended networks through which cast-offs are exchanged and recycled commodities are traded, working closely with translators in the various markets I visited.

At first, much of my everyday living and working involved talking to neighbors and extending my contacts through them to other housing societies in order to understand the domestic economy of clothing, and how and why people got rid of cloth. I found that most of my neighbors admitted to an almost complete lack of knowledge about the market for used clothing and what exactly happened to their cast-offs. I initially had to operate in several seemingly discrete spheres at once, and often could only link them up through a chance article in a newspaper, a surprise meeting in a back office, or a lucky search online. The disconnected nature of my research reflects the social distance of the various communities from each other and their lack of knowledge and awareness of each other. But they are linked by the unexpected movements of people and objects, as will become clear.

In this ever-changing environment, residents strive to create new identities, and communities develop. As one founding resident put it, "These housing cooperatives are generally formed when like-minded people get together." The founders often work together, and the development may be underwritten by their employers, hence Oxford University Press Apartments, UNESCO Apartments, Pharma Apartments, and so on. Occasionally language groups or regional associations form the basis of a housing society, as was the case with Mithila Apartments, and religious affiliations may do so as well. The Progressive is perceived by its occupants as being rather special, and is largely made up of academics, writers, journalists, artists, lawyers, and other professionals. The founders either worked in such fields or were close family members of those who did, and all those subsequently wanting to join had to be vetted. About 50–60 percent of the apartments are owner-occupied, the rest being let to tenants on short leases. Some of the tenants managed to stay in the cooperative for several years by moving annually within it, and eventually a few were able to buy in.

The majority of the occupants were educated, upper-caste Hindus from moderately well-off backgrounds; they tended toward left-thinking and socialist principles, living thrifty lives and eschewing extravagance. Several couples had inter-caste marriages and many different religions were represented. There were approximately four Muslim families, a few Christians (both Syrian Catholics from Kerala and Protestants from the northeast hill states), Sikhs, a Parsi household, a Chinese family, and several mixed marriages in which the partners had different religions. On a whole cosmopolitan in outlook, the residents came from all over India, bringing with them regional languages, cooking, dress styles, and social and ritual customs. Many had migrated from the south, west, and east of India for work opportunities, whilst others came from established Delhi and north Indian families. Many had also spent periods living and working abroad, or had close family outside India. Personal circumstances such as internal migration, the decline in large joint households and the growth of smaller nuclear families, or retirement from government jobs, combined with rising property prices, had led them to move to the new Trans-Yamuna settlement rather than to south Delhi or the older colonial administrative center in the north, Civil Lines. This move could constitute a risky, but necessary, bid to maintain their lifestyles as far as possible, among like-minded neighbors, despite the unappealing surroundings. Leafier suburbs in Delhi were often too expensive for those on public salaries, while the rapidly growing housing developments in Trans-Yamuna offered a new opportunity to form a community, although they often developed ahead of social infrastructure or environmental planning. The mushrooming of these housing developments is a reflection of the rapid growth of the middle classes in Delhi and the social juxtapositioning of families rich in socio-economic wealth and those possessing high symbolic capital.

During the year, neighboring societies were constantly hosting regional and religious festivals both day and night, as well as the occasional

Jagaran, where devotees come to eat and listen to Hindu devotional music. In addition, families hosting wedding parties would erect large canopies over the gateways to their apartment complex and monumental marquees in the car parks and garden areas inside, with wedding bands, discos, and Hindi film music broadcast on loudspeakers at deafening decibels all night. Streams of guests crowded the roads at night in fabulous, glittery attire, sporting their best silks and gold jewelry, as they climbed in and out of new Maruti cars and taxis.[2]

Members of the Progressive do celebrate such occasions with their families and friends, but usually the events are altogether quieter affairs, characterized by eating together, chatting with friends, and dressing in a simple manner, in keeping with the ethos of restraint and appropriately thrifty behavior fostered in the society. The major festivals of most religions were marked, including Diwali and Id, and everyone joined in the water fight at Holi, the festival of color. Other festivals were celebrated by particular groups, such as Janam Ashtami (Krishna's birthday) and Karva Chauth (the north Indian festival where married women worship their husbands). But the housing society's management committee (MC) did not encourage large-scale "religious parties," which were characterized by some as an opportunity to show off, dress up to excess, and disturb the peace with amplified music. Significantly, several occupants I talked to repeated that "that culture is not here," that the "new rich and the showy business-class habits are not welcome," that "they can stay here [in the Progressive] if they adapt themselves to our culture. No one will stop them behaving in their own manner, but they are odd men out—and no one wants to be the odd one out."

As one academic member of the MC explained, a small number of the Progressive's residents are from the business class, but most are "Brahmin by birth or by profession, meaning that they earn their livelihood through intellectual pursuits." Ostensibly, residents are not discriminated against on the grounds of race, caste, or religion, and are free to choose their own lifestyle. Yet, with a strong dose of irony, he referred to M. C. Srinivas's concept of "Sanskritization" (1956), implying that by moving into the Progressive one can improve oneself socially—one can "move up in caste."[3]

In the early days, new residents would help each other out, filling up water tanks, sharing lifts to work, or designing the gardens. They made a conscious effort to celebrate each other's festivals and have communal lunches that they prepared together, and many families formed close bonds of friendship. Recent changes in the law mean that the society can no longer prevent people from buying and selling flats in the cooperative, although the MC has the power to expel members. One story recounted to me had already acquired the characteristics of an urban myth. It concerns the first member who tried to sell his flat without involving the MC. He had sold to the highest bidder, a family from the business class whom nobody knew. When the MC found out, they physically barred the gates and the new owners were left standing with their possessions outside. Although high-handed and illegal, these tactics had the desired effect; apparently the message spread like wildfire to prospective buyers, preventing those without introductions from daring to buy through fear of suffering such embarrassment. The MC proposed an informal system whereby prospective buyers should be introduced to the society by at least three original members, and that, in consideration of the large profits being realized, 15 percent of the apartment's original purchase price should be given to the society's coffers for investment. According to a long-term resident, the society was now getting the "right" people and plenty of money.

A more pressing anxiety among owners is caused by the need to constantly keep an eye on the long-term social development of the Progressive. Most owners are not very rich, but a few live in other properties they own or in houses that come with their jobs until they retire. Therefore for some the apartment in the Progressive is an investment, while others may choose to move in at retirement. One resident couple were worried that the whole area would become a slum in a few years' time, and sometimes they thought about moving out. Many people took a significant risk in moving to Trans-Yamuna, and their social standing is by no means secured in this fluid environment. This constant worry and tension affects their evaluation of the physical space within the society. Some areas within the Progressive are considered better kept than others—an elderly couple remarked that the stairwell where I lived had "good people" and was tidily kept compared to others, where the occupants were less "civic-minded."

Clothing serves an important function for these residents as they move through the society throughout the day, revealing the ambiguous nature of the boundaries of both group and individual identity. Underlying the way in which relationships are conceived and enacted within these places is the important dichotomy of "inside" and "outside," where "inside" is fundamentally equated with home, family, and the domestic unit.

The Progressive is arranged into four buildings around a central courtyard, and is contained within a perimeter fence lined with trees and encircled by a concrete path. The front gates open onto the main road. Every flat has at least one balcony facing the common courtyard and there is a feeling of openness, greenery, and community living. In addition, flats have balconies at the rear, facing the perimeter, allowing for some privacy. The ground plans, staggered façades, and tiered stories of the buildings mean that no flats directly overlook their neighbors. The four buildings house roughly a hundred flats of varying sizes, from small one-bedroom apartments to large four-bedroom maisonettes. The average flat houses a semi-nuclear family of four or five people, though there is a wide variety of living arrangements, from young couples on their own or retired couples, to those with children or an elderly parent at home, to joint family arrangements with grandparents, married children (typically one or more sons and daughters-in-law), and their children all in one large flat. There was only one instance of young professionals sharing accommodation while I was there.

When residents speak of the Progressive collectively as "inside," they are including their apartments in a whole that also encompasses the social space of the inner gardens which everyone walks through, enjoys, and stops for a chat in on their way to and from the "outside" beyond the gates. This inner garden is well kept, with green grass, clipped low hedges, and brightly colored flowerbeds. Towering climbers ascend the buildings; in full bloom throughout early summer, they attract small tropical birds and butterflies. People walking through this space are well dressed, and tend to behave courteously: this is the society's "front," with a reputation and character to keep (Goffman 1971). The maintenance of safety, of the barriers between inside and outside, is of utmost importance—the Progressive is perceived by its residents as a haven, even by some as an oasis of calm and green in a city gone mad, and the threat of disorder and disharmony is ever feared.

By contrast, the outer perimeter of the society is rather bleak, and is marked by a concrete footpath. Rear balconies hold drying racks full of tatty underwear, and householders casually appear on them in states of domestic undress. The rubbish is put out and collected every day from each flat, and a sweeper empties it into a cycle cart pushed around the perimeter walkway. Unwanted furniture and coolers pile up, appliances wait to be mended. On each side of the Progressive, beyond its perimeter fence, are rather ugly neighboring housing complexes, both suffering from neglect, with their concrete crumbling. The rear perimeter is opposed to the public face of the society, and can be entered by a back gate from a piece of undeveloped public land, where municipal rubbish is collected.

Many residents commented on the fact that before the economic liberalization of the early 1990s, many domestic goods purchased in the bazaar were basic raw materials, to be processed back at home. For example, few processed foods were available, and packaging was minimal; raw ingredients were wrapped in small packets made from old newspaper, which is now used only by food sellers in the street; shopping is now often carried home in thin stripy blue or pink plastic bags. But although convenience foods are increasingly available, raw ingredients are still largely transformed into meals in the kitchen through culinary skill and hard work. Thriftiness prevails, and it is often said that nothing is wasted. The same philosophy applies to most material goods in the home: they do not become "rubbish" but are recycled for the value of their raw materials at least, or sometimes for their functional utility, as are bottles. Anything that cannot be used and reused within the home can be sold back to the *kabariwale*, the dealers who come to the gates of the society every weekend. Arriving with a set of scales, they buy up and recycle old newspapers and scrap metal, and may even take broken furniture and electrical appliances.

Therefore "rubbish" mainly consists of food waste such as vegetable peelings and, nowadays, commercial packaging, whose volume is outstripping households' ability to reuse it domestically, but which is so far not collected by the *kabariwale*. The amount of garbage collected from contemporary urban living spaces is increasing, partly because households have fewer opportunities to process and recycle their own rubbish. Households do not sort rubbish into organic and recyclable categories; everything not used or saved in the home is simply left outside on the doorstep every morning.[4] What little is thrown away is first sorted by the man who collects it from the doorstep, who removes anything he spots which has a resale value, such as food tins and plastic water bottles. These can be sold to small middlemen specializing in recyclable commodities bought by weight. By leaving them on the doorstep, residents eliminate the need to find somewhere to sell them, instead tacitly giving them away as gifts to the sweepers, whom they know will earn something from them. This leaving of rubbish is strategic rather than habitual for most residents; many people will stockpile plastic, paper, and cardboard packaging for months and years in the hope that they will be able to use it up themselves, a typically thrifty form of behavior orientated toward the household. The remainder is taken to the municipal dumping ground behind the society and sorted by the ragpickers; the food waste is fed to urban pigs every evening, which are led along the back streets to the dumping ground by their owners. Wheeling kites dot the skies, a sign of natural recycling in an urban environment.

Daily Life

Observing the clothing of all the various occupants of the society during the day and relating it to the contents of their wardrobes not only reveals the process of self-making, but also unfolds a map of social interaction through the strategies employed in its performative display. This focus reveals cloth to be an interweaving medium that hangs down through social hierarchies, identifying vertical semi-permeable columns of relationships; drapes across social layers, covering and concealing inequalities; and envelops generations, constituting them as family insiders.

The day starts early for most households, typically before dawn. The early mornings are the coolest, most pleasant part of the day during the hot summers; windows can be opened for fresh air, before being closed to maximize the efficiency of the ubiquitous "Desert Coolers," or air conditioning systems, and to vainly try to keep out the carpet of red Rajasthani dust that settles over every surface. Yet even during the depths of winter, when tiled floors are icy to the feet and thick yellow smog envelops the outside world, families take advantage of the hours before work and school. Women rise early to take a bath and dress before beginning household chores, including the daily cleaning of the flat. For those who are religious, prayers must be said and offerings made to the household shrines. There is tea to make and drink, breakfast to cook and eat, and lunchboxes to prepare. Daily deliveries begin soon after dawn—the newspaper boy throws furled papers up onto the balconies, and the milkman delivers fresh milk, ladled out of churns on the doorstep.

Taking a walk around the periphery of the society about 6:00 AM, one meets a few people up and about, has a word with them while they are out on their balconies, and hears varied sounds of cooking, washing, or hawking and spitting through open windows. Most people are in the private realm and are not yet properly dressed—men wear *dhotis* wrapped around them as they go in and out of the bathroom, or loose old *kurta py-jamas* (tunic and trousers) around the home. Women often wear the popular European-style nightie, reaching to the ground and with sleeves and a gathered yoke, with a shawl around their shoulders, or alternatively an old sari or *kurta* with non-matching *salwar* (trousers) and *dupatta* (scarf). Occasionally a man might be out walking, but more likely a few women will be taking a stroll, chatting in twos or threes by a ground-floor balcony or terrace with neighbors, or more earnestly pacing the perimeter in a bid to get fit or to lose a little weight.

Plant pots and flowerbeds around the ground-floor flats and balconies are tended and watered; a couple of women cultivate special flowers and leaves that they pick for daily offerings at the shrines. Some practitioners may be doing yoga exercises, some people are listening to or chanting along with tapes of Hindu devotional songs, and musical instruments are being practiced, newspapers read, and homework and study caught up with. In the school holidays young children are often out early on their bicycles and tricycles, yelling up at each other's balconies, urging friends to come and play. But in term time they are escorted out to the street to catch early buses to school, wearing European uniforms and clutching Disney lunchboxes and water bottles.

By the front gate, the *chowkidars* (guards) have changed shift, and are checking all who come in and out. During the night the fenced society is patrolled by the guards tapping their *lathis* (long bamboo sticks) and whistling to each other, and only the occupants are allowed inside. They wear tatty, but smartly presented, military-style uniforms with their sticks strapped to their leather belts. And every morning a contingent of three or four men are busy washing, drying, and polishing all the cars, using old cloth rags. They come to work for an hour or two every morning, in torn T-shirts and old *dhotis* or trousers. By the single tap for potable water, a queue of residents (both women and men) and maids wait to fill their containers for the day.

A little later, by 8:00 A M, the older schoolchildren have also boarded buses, and most people who are working in offices are leaving as well. The relaxed air of the very early morning around the back of the flats is replaced for a short while by a flurry of activity in the central arena at the front, as people pass through the public gardens and out to their cars or buses. Men are now usually wearing Western-style trousers and shirts, and Indian jackets if need be. Some of the academics wear a formal long *kurta* instead of a shirt with their trousers, or even *kurta pyjamas,* woven from cotton or perhaps *khadi* (homespun), in white or natural shades, always clean and pressed. Women going out to work may be wearing saris or smart "Punjabi" suits (*salwar kamiz*), which consist of a long shirt or tunic-dress (a *kamiz* or *kurta*), baggy trousers fitted around the ankles (*salwar* or *churidar*), and a matching scarf (*dupatta* or *chunni*).[5] Although there is an enormous variety of cloth, cut, and color within these basic styles, the smartness, neatness, and harmony of an outfit are of paramount importance. Younger men and women going to college may be wearing Western-style jeans and shirts or suits or, for girls, ankle-length longer skirts and blouses.[6]

As the rush passes, the character of the area changes once more, as those left at home are generally either housewives, academics, or artists working at home, or the retired and elderly. The bevy of maids that work for most of the occupants have been arriving and now the process of dusting, sweeping, clothes washing, and cooking begins in earnest. The sounds of buckets of water being filled and clothes being washed and pounded with wooden beaters (*danda*) echo around the buildings. All the floors are mopped, and bedcovers and throws are shaken out and aired—soon the whole society, back and front, is festooned from top to bottom with brightly colored saris, bedsheets, and clothing of every description hung out to air or dry. Balconies are filled with colors and patterns that hang in long swathes, tumbling down from floor to floor, resting on bushes and trees and flipped across every available piece of line. Loud hissing sounds seem to threaten explosions as scores of pressure cookers prepare the midday meal.

The maids wash floors, dishes, and sometimes clothes, and may help with cooking. They often work for several people in the society, moving from flat to flat, and if they cannot come one day, they may send a sister-in-law or daughter in their place. Therefore there are networks of maids overlapping networks of women in the society, creating links along which things may be lent and both written messages and informal gossip are passed. Most women have some help, and a few have live-in servants. All the maids wear cast-off saris or suits, usually given to them by their employers, but often their blouse (*choli*) does not match their sari, or their *dupatta* and *salwar* do not match their *kurta*. As a result of both their origin and the nature of their wearer's work, the clothes are generally tattier and dirtier by midday, when most leave to go home to their families. If the maid's employer is also at home in the morning, overseeing and often herself doing chores, she is probably wearing older, casual clothes.

As the morning progresses, the occupants come and go, visiting each other in their homes, going to the fruit and vegetable stalls situated just outside the gate, or walking a hundred yards up the road to the local market. Most middle-class Delhiites usually dress semi-formally for such daily social contacts, donning clean and presentable suits or saris before leaving the house. What a woman decides to wear will depend upon the time of day, the season, where she is going and who she may meet, any particular religious occasion which she wishes to mark, and her own personal taste and style. Rarely will she be seen out in something that she has just been cooking in, that has a tear or a dirty fall,[7] or that does not match, and if she is wearing a suit she will not omit the *dupatta,* even just to "nip out to the shops," unless it is unavoidable. For appointments farther afield and trips to town they often use the local taxi firms that cluster around a telephone rigged on a lamp post, where the drivers shelter under tent awnings. A throng of cycle and auto rickshaws wait for business outside the gate, and all morning a stream of private charter buses take passengers throughout the city. The local city buses that run from the depot at the other side of the market are far less comfortable, being hotter and more crowded, but cheaper; however, few of the residents used them. The mode of transport is as important to the choice of what to wear as is the social character of the destination in such a dirty and polluted city as Delhi, and clothing's cost, fabrics, color, and ease of maintenance are all taken into consideration.

During the day, the regular workers, such as the plumber and the electrician, are signed in to the society, in addition to the maids. Jasbir is the contracted electrician; he arrives every day by scooter from the slum colony down the road to fix whatever problems occur, add new sockets, connect modems, and overhaul cooling systems. He is paid a retainer for doing so and is a semi-skilled worker who regularly goes in and out of everyone's apartment. He is reasonably tidily dressed and wears Western trousers and a shirt to work, never a *dhoti* or a *kurta*. The society also employs an elderly gardener, or *mali baba,* who wears an old shirt, a ragged *dhoti,* and a turban. He trims the hedges and waters the trees and plants, and prunes under the guidance of some of the more knowledgeable inhabitants. Sometimes he brings his grandson, in old clothes that are far too big for him—his trouser legs and sleeves are rolled up around his knees and elbows. Often he is helped by a young sweeper or *kuruwala* named Suman, who clears up the hedge cuttings and is always dressed in worn trousers and an old shirt. The gardener, his grandson, and the sweeper are usually barefoot.

The electrician and the plumber are engaged in servicing the requirements of modern middle-class living: they don their semi-presentable Western-style clothing and accoutrements during the day, and use their technical skills to move between suppliers and often anxious, yet defenseless, homeowners. Their clothing helps them construct their authority in the domestic arena through these ongoing performances. In contrast, the gardener and the sweepers rarely go inside homes, but are concerned with the maintenance of the public sphere and the removal of waste. These workers are of the lowest social status, always making deferential full *namaste*s or obeisances in the traditional manner of greeting to everyone passing, and wearing the most tatty cast-offs that have been donated by some of the residents.

The regular workers at the Progressive are known and trusted by most of the inhabitants and the guards. The guards themselves are provided by a local agency, but they often work at the Progressive for several months or even years, and get to know the residents, guests, and local workers. They distribute mail, pass on messages, and help with small chores. When asked, they call in the barefoot delivery boys who bring black-market bottled gas, piled precariously high on bicycles. They can also contact suppliers such as the *cablewala,* who installs cable TV and collects the monthly rental fees. As the purveyor of new technology and representative of a large company, he is always dressed in a dapper fashion, wearing stylish Western clothes such as a bomber jacket, a checked shirt, old jeans, and cowboy boots that are at least three sizes too large. All look rather worn and ill-fitting, but are presented in the nature of a cowboy performance; his stature increases as he tries to fill out the outsize garments and clomps across the stone floors, thumbs in his belt loops.

Outside the gates of the Progressive, the *presswala,* who does the ironing, sets up his stand, with a wooden bench under the shade of a tree and a large iron full of glowing embers. Because much middle-class Indian clothing and domestic cloth is pure cotton, he is constantly busy. His children go from door to door in the mornings collecting bundles of ironing tied up with an old torn piece of sari or *dhoti.* In the evenings they return the clothes, charging Rs 1 or 2 for each item. His children are grown now, but they played with the children living in the society for years, and they are always dressed in a mixture of ill-fitting, tatty clothes that are Indian in style rather than Western. He himself wears a T-shirt and a *dhoti,* and his wife a sari, as do all the other *presswale* in the street. They are perhaps just a little more smartly dressed than the fruit and vegetable sellers who line the roadside up to the local market, sitting on the edge of the storm drains in torn old *dhotis* and T-shirts.

About 2:00 PM, some teachers, lecturers, and other half-day workers come back to have lunch, along with younger children whose school has finished. The society has a peaceful air on the outside, especially during the heat of the summer, when lunch is eaten, siestas taken, and reading and homework done. The sweepers, the gardener, and the *presswala*'s children may be sleeping on the concrete paths in the shady spaces below the apartment blocks, or huddled near the electrician's cupboard under the stairwell in winter. As the afternoon progresses into early evening, the gates are constantly busy once more, with residents returning home from work, school, and college, and some maids returning to prepare the evening meal. The bustling activity of the early morning is replaced by a more leisurely air of sociability: friends and colleagues gossip by the gates and call down to one another from balconies, maids and young girls push babies around the perimeter, older women take an evening stroll, young children ride bikes and roller-skate into each other while older ones compete at cricket and badminton, and teenage girls chatter and giggle on benches. Sometimes women sit out in the central grassy area chatting, reading books, or playing with their children. Visitors call on each other for a chat, a favor, a cup of tea, or a drink and a gossip.

On returning home, people bathe again, and change into more casual clothes if they are now inside the complex for the evening—softer cotton suits or older saris for women, *kurta pyjamas* or tracksuit bottoms and T-shirts for men, and jeans and T-shirts for teenagers. However, many residents will be going out again to meetings, supper invitations, or events such as religious festivals, birthday parties, and weddings. For such things, everyone smartens up and dresses appropriately. Weddings are the time to dress in one's finest Indian outfits, the more so if one is closely related to the bride or groom, as it is important to display one's status and not let anybody in the family down. Children wear bright party clothes, with very young girls in frilly dresses or sparkly *lehenga choli* (long skirts and blouses). Men don smart suits, *kurta pyjamas,* or more rarely *dhotis.* Women should be dazzling, in bright silk "heavy" saris (which are woven and embellished with gold and silver threads and can be literally heavy and exhausting to wear, especially for hours-long weddings), gold jewelry (earrings, necklaces, bangles, and rings), and more make-up. Most such celebrations take place in wintertime, and outfits are topped with finely embroidered shawls to match.[8]

For more informal parties and suppers, men and children wear a mixture of smart Western casual wear and Indian styles, and women wear cotton or silk saris and decorated suits. In contrast, trips to the large, air-conditioned, English-language cinemas in south Delhi often call for the whole family to don its trendiest Western gear, though the women may not.

On turning in for the night, the society's residents revert to the softer, looser clothing seen in the early mornings—nighties or old saris without blouses for women; pyjamas, *dhotis,* or underwear for men; and perhaps cartoon pyjamas and nighties for children. The contractors, maids, and sweepers all return home to their slum colonies in neighboring areas. They too revert to simpler, non-Western clothes—saris, *dhotis,* and pyjamas. The *chowkidars* sit sleepily in their second-hand uniforms by the gates, while outside the rickshaw pullers contort themselves across bicycle seats and itinerants curl up on the pavement and in doorways. For them there are no appropriate changes of dress as the day progresses; opportunity and opportunism are the immediate influences on their sartorial calendars. In between, their clothes and bodies remain unwashed, and they wake up to a new day a little grimier than when they went to sleep.

The Archaeology of the Wardrobe

Family homes in the Progressive had relatively few personal possessions and decorative pieces on display compared to family homes in south

Delhi, for example. A couple of pictures on the wall, a piece of framed cloth, the odd photograph, a carved animal, or an occasional table lamp was often all that was evident in living rooms; rarely would one see a whole collection of objects or images together. Furniture was often fairly new, and in a modern Western style except for cushioned divans. Books were usually shelved in studies doubling as guest rooms, together with general clutter. Kitchens were modern but simple, and had not yet been converted into the Western style with fitted cupboards. They consisted of a marble worktop with a fitted sink, a two-burner gas stove, and a shelf running around the wall housing a gleaming range of steel utensils. Bedrooms had double beds, large cupboards, and a range of boxes and cases stuffed with cloth and clothing. The contemporary popularity of overhead strip lighting often gave rooms a bright, harsh atmosphere, made even less forgiving by the unadorned, pale blue, lime-washed walls. Often the only softening of this atmosphere was the use of cloth; a multitude of colors and textures are used as coverings for divans, sofas, and chairs, or as runners on top of tables, sideboards, or the prominently displayed refrigerator.

Women are usually the most concerned with cloth in the household, as the main organizers of purchases on behalf of family members, principal receivers of cloth gifts, and everyday managers of wardrobes.[9] In order to find out how clothes are valued in contemporary urban India, I initially conducted around thirty semi-formal interviews during which I took notes and photographs. I spent hours trawling through women's wardrobes with them, sifting through trunks and hearing stories of each sari, suit, and surreptitious pair of jeans, listening to tales of mothers and mothers-in-law, grandmothers, weddings, favorite aunts, and irritating cousins.

These interviews were a kind of intensive "wardrobe archaeology" of clothing *in situ*. I would usually be invited to sit on the bed whilst the woman opened up her cupboards and trunks, suitcases and boxes, pulling out clothes and recounting their histories and associations, and commenting on her particular likes and dislikes. I noted how clothes were stored, in what types of furniture, and where in the house; whether the summer or winter wardrobe was being shown; and the extent of material stored elsewhere. I was interested in the level of the women's knowledge about the construction and decoration of textiles and in their individual preferences for colors, styles, and patterns. I asked how much of the contents of an informant's wardrobe had been gifted to her and how much she had purchased for herself. I then focused on investigating strategies of keeping and disposing of clothing during the owner's lifetime.

Often I would prompt stories by questioning, but usually the women's own sense of priority and importance, as well as the arrangement of the wardrobes and their contents, dictated the order in which clothing was produced and the narratives that accompanied its showing. Such an organic mode of excavation frequently elicited the women's own mental and physical structuring of their wardrobes, unraveled associations and connections between items, and permitted the spontaneous expression of feelings, emotions, and deeply held beliefs woven into garments. Sometimes children, spouses, and family members would join in, adding their own observations and insights.

Twenty-three of the women I met in and around the Progressive are quoted in this book. The names of these women and all other informants have been changed. A few details about their family backgrounds, regional origins, and education have been included as the women are first introduced; specific information has been edited to preserve anonymity without changing its contextual meaning. Most of the women referred to here were upper-caste and many were Brahmin, but in fact nearly all of them were either the children of "mixed" marriages or had themselves made such alliances, whether the differences were regional, religious, or of caste background. Hence, as many women discussed the impact of mixed heritages upon the symbolic presence or absence of traditional clothing, the style of wedding trousseaux, and what was deemed acceptable to wear in the different cultures in which they had lived, the narration of their wardrobes was always highly individual and contextual. The extent to which they were able to be assertive in their choices of what to acquire, what to keep, and what to get rid of reflects the fact that, as a technology of the self, the management of the wardrobe can be heavily implicated in domestic politics.

Buying Second-hand Clothes

The first intensive phase of fieldwork largely focused upon the middle-class wardrobe in a domestic setting; at the same time I also started traveling across the city, visiting local bazaars, boutiques, the antiques market and modern retail complexes, the weekly suburban market (*haat*), and especially the large Sunday market, the Chor Bazaar, at the Red Fort (Lal Qila). I was interested in the categorization of clothing, and in what value a piece of cloth or clothing has in different contexts. How is its value constituted by how it was obtained, where it was sold or resold, and the consumers for whom it is intended? For it was already proving difficult to correlate what the relatively conservative middle-class women I knew were telling me about the origins and fate of their own clothes (both Western and Indian) with the hidden-away markets, overflowing with used and reused clothes, that I found across the city. These transformations of value took place at every level, in street markets as much in elite boutiques. Uncovering the lives of a sari involved following the unfolding topography of the city and the often covert relationships between socially disparate spaces.

Suburban *Haate:* The Weekly Evening Market

The most ubiquitous markets where I started working were the weekly *haate* found in middle- and lower-class suburbs across the capital, which offer the shopper a huge array of food, clothes, and cheap household goods. Both middle-class residents and their domestic workers and service providers regularly shop in the local *haat.* Carts are laden with seasonal fruit and vegetables, while other traders spread tarpaulins on the ground for the piles of fresh herbs and spices. Many dealers set up stalls selling plastic bangles and hair ornaments, cheap steel kitchen utensils, plastic bowls and tubs, toys, religious posters, or music cassettes. Several dealers sell cloth products for the home, such as mass-produced printed bed sheets, towels, tea towels, durries (woven rugs), oven mitts, and so on. In addition there are the clothes traders, selling poor-quality staples such as *choli* cloths, petticoats, saris, *lunghis* (men's wraps) and trimmings, jeans, shirts, and Punjabi suits to the less well-off, who rarely buy things in more formal shops. Other garments on sale change with the seasons: warm knitwear, shawls, and blankets in the winter and lighter synthetic nighties, suits, and saris in the summer.

There is evidence of resourceful reuse and recycling in every category of object and material in the market. Household goods are often handmade from faulty factory-printed sheets of tin and plastic, such as the metal food graters showing repeated motifs of sardines swimming in a sea of tomato sauce, the flimsy plastic kites adorned with multiple images of a hunched

Sai Baba leaning on his stick, and the hand mirrors mounted on cardboard sleeves from A. R. Rahman's latest *filmi* soundtrack.[10] Buckets are often made from mixtures of recycled plastics, bright candy colors swirling on a white background like raspberry-ripple ice cream. Yet the origins of the various articles of cloth and clothing are much harder to decipher. Closer investigation reveals the boundaries of the categories of "new" and "used" to be highly permeable.

Although my research focused on used Indian clothing, aiming to find out what it could become and what determined those options, it was often difficult at the beginning to tell exactly what it was that was being sold. I was initially looking out for any obviously recycled cloth products, second-hand clothes, and reused cloth, perhaps similar to the sari products I had seen everywhere in the UK. But the look of cloth and clothing was often deceptive, even after feeling the cloth, smelling it, turning it over and looking at its construction as well as its style, colors, and designs. The term "recycling" can mean reusing both the by-products of production and post-consumer waste, and it quickly became clear that the strategies employed by manufacturers, traders, and retailers to transform "waste" (or, perhaps, "leftover") textile materials of all kinds into saleable commodities were extremely complex.

Many of the typically Indian clothes in the markets, such as saris, suits, sari blouses (*cholis*), petticoats, and *lunghis,* were new, manufactured from dyed or printed machine-woven cottons and synthetics. Intermingled with the dealers in these were several stalls selling Western-style clothing for men (shirts and T-shirts, jeans, and trousers) or for girls, or women's indoor wear, such as maxi-dresses, skirts, and blouses. The girls' and women's clothes also tended to be in Western-style colors and printed patterns. It is unclear how many of these clothes were in fact originally manufactured for the Indian market, as, when asked, most traders described them as "export surplus," a ubiquitous category that could include all types of clothing in every stall. In fact, not once did a trader ever claim that such Western-style clothing was Indian. "Export surplus" normally refers to clothes manufactured in India for export, to designs and specifications created by Western companies, but which have been rejected for poor quality (seconds) or for arriving too late at the docks for an export deadline, or which are extras, manufactured to cover an expected percentage of faults in a production run but ultimately not needed. These clothes are diverted to the local market at a cut price.

The epitome of such export surplus is the Western-labeled goods sold in the market off Janpath, in the city center, and in Sarojini Nagar, which was frequented by the local middle-class teenage girls. The phrase "export surplus" endows the clothing with the blessing of Western design and appeal, while concealing the faults usually found in seconds—shoppers need to be discerning to establish each garment's credentials. However, the suburban markets rarely sell "surplus" garments from upmarket shops like French Connection or Monsoon, or even the viscose blouses and skirts from cheaper branded stores.

The category "export surplus" expands even further to include a whole range of clothing manufactured in India for sale on the domestic market. These products utilize the rolls of cloth left over after a print run once the required number of jeans, shirts, or dresses have been made up for a Western order. There is a dedicated "export surplus" cloth market, Shanti Mohalla, part of Gandhi Nagar in Trans-Yamuna. One dealer outlined the system: were a manufacturer to contract to print 60,000 meters of cloth for a foreign order, he might actually print 62,000 to cover any faults. The resulting surplus is sold off at half price to dealers, who in turn split the large rolls and sell them off again to small scale manufacturers, a few meters at a time. Although apparently most foreign companies specify that printing screens be destroyed after a print run to try to stop local printers repeating runs or copying designs, the surplus material that remains after the original print run, known as "cabbage" in the West, constitutes a significant resource for the local market and guarantees an extra bonus for the manufacturer.

This surplus material is usually fashioned into Western clothing—denim into jeans, striped cottons into men's shirts, flowery synthetics into women's skirts. Standards of production are far lower than those demanded by foreign companies, though many have labels, or embroidered logos on pockets or shirt fronts, imitating well-known Western brands. Even spurious brands indicate "Made in USA." In fact, it is relatively easy to buy rolls of cloth labels made up to say anything required by the latest fashion, often misspelt or nonsensical.

It is beyond the scope of this text to investigate the Indian manufacture of copycat branded goods and the internal and external black markets in fakes and forgeries. The example serves the purpose of providing a paradigm of acceptability in the market, highlighting the complex origins of cloth and the way in which fabrics and styles are habitually manipulated to increase its desirability. Whether or not the copycat garment actually succeeds in convincing the buyer or wearer that it is an authentic product of the brand is a moot point—that is, in all likelihood, not the primary intention of the label or logo in a society where few read English, if they can read at all. What is important is that the overall image conveyed by wearing the garment look right and have the desired effect, usually ensuring that the performance passes as suitably modern, fashionable, or Western. Indian clothing is generally not branded, though styles can be identified through a multiplicity of other codes, so it is necessary for the fabric of these Western clothes to be branded through labels and text, whatever their content. Status-conscious upper-middle-class buyers who do purchase such goods are careful to avoid looking ridiculous by wearing poorly pirated logos, but occasionally they are caught out. But, with the focus on applied text, the reuse of surplus cloth is itself concealed. It has been diverted from its path as an object for export; unwanted, it was classified as waste material by the company. It is the material qualities, the colors and designs, of the cloth that were valued by the small-scale producer of the Indian garment, enabling him to mimic the "real" thing.

A related category of materials could also be included in "export surplus": the scraps and defective pieces that are the by-products of the manufacturing process. They are not recycled only in the export surplus market of India; the practice is an established part of the modern textile manufacturing industry worldwide. The particular uses of these by-products depend largely upon the fibers used, the nature of the cloth, and the available technology, and they are too numerous to detail here. Yet the trajectory of cloth, from factory to petty trader to domestic market, also allows for a high degree of innovation in the creation of new products. One market trader was selling new sleeveless jackets made from cut-pieces, ends of the roll of wool-and-synthetic-blend suiting from the factory: he bought the cloth in summer when it was cheaper, and made them to sell in winter for Rs 125.[11] The *chindi* (rags) and *katran* (scraps) have their own specialized markets where small-scale dealers buy up waste directly from the factories. In mill towns such as Ahmedabad, spinning, weaving, cutting, and tailoring all create materials that are sold off and reused in a variety of ways by the informal sector (see, e.g., Das and Bhargar 1987).[12] At the Mangolpuri *katran* market on the far outskirts of Delhi, families buy sacks of scraps, each weighing from two hundred to a thousand kilos, sort them by color and size, and sell them for a few paise per kilo.[13] The long thin rags are used for durries,[14] longer rectangles go to the machining industries and car trade for polishing, and satin pieces become hair ties and pocket linings; the tiniest scraps are used as mattress stuffing.

On a walk through the local markets in Delhi one can find these scraps reused in a variety of domestic products: wash bags, hair scrunchies, shoulder bags, and patchwork clothing. Dresses are made up with contrasting sleeves, bodices, skirt panels, and facings, with clashing fabrics mixed and matched in wild abandon. Such hybrid garments cannot be considered wholly Western in design: basic elements appear to be derived from Western styles, and yet the clothes are manufactured in India for an Indian market. Patchwork, quilts, and the popular durrie rugs: all these technologies and products are based upon domestic practices of thrift that are translated into the industrial commercial sphere and exploited by the entrepreneurial middleman, and their existence helps impoverished rag pickers eke out a living.

Since most of the products that are not "traditional" Indian styles are derived from the export surplus market in some way, their materials have already been classified as unwanted waste, cast off and recycled before they ever reached the end consumer. Such products are not inauspicious and display resourcefulness or ingenuity, which have positive moral values. Additionally, as they have not been worn or used previously, they pose no risk of pollution to orthodox Hindus, as they might if they had originated from someone of a lower caste or in a ritually polluted state. Many of the local middle classes would readily buy a printed bed sheet, blanket, durrie rug, or piece of cloth from the market for an everyday purpose. These would not be valued as highly as handmade pieces from a craft fair, charity event, or Fair Trade organization, but they are often attractive and cheap, and therefore a thrifty purchase.

Alongside these cloth goods in the local markets are clothes that are openly acknowledged as used. A few Waghri people are occasionally seen selling used Indian clothing, but in middle-class areas the trade is less successful; everyone knows that the clothes are second-hand, and therefore only the poor will openly buy them. More choice is to be found in the specialized weekend markets, out of sight from neighbors. One or two stall holders will be selling Western used clothing, branded goods that have been manufactured across the globe, bought, and then cast out of

Western wardrobes. These are technically illegal imports, although container loads are smuggled into India every year. Large bales are broken down and sold off to the small-scale "footpath" merchants (*patriwale*) by the kilo. In summer they predominantly sell trousers and shirts, while in winter jackets, sweaters, scarves, and woolly hats are displayed. Many of the winter goods are not easily available new to the poor—thick winter coats and anoraks are not typical Indian products and are often imported, and thus very expensive when new. Although they may look worn, they are often in desirable styles and colors. They undoubtedly provide a necessary resource for the local workers and small-time traders, but would be too expensive for the absolutely impoverished.

But the nature of the goods in the market is often difficult to discern: it is a tricky place for the unaware, which can be an advantage for the less scrupulous bargain-seeker. Some second-hand clothing in good condition is passed off as export surplus by the traders, and indeed export surplus garments can be similar in design to used clothing and equally grubby. This similarity can be used to advantage by a shopper, who can claim it is a new garment to others, however suspicious she may be herself about its origins. When I spoke to my middle-class neighbors in the Progressive, it was clear that they would never openly admit to buying any of the used clothing in the market from whatever source.[15]

Sunday Markets: The Red Fort

If the local markets present a confusion of clothing, and the elite boutiques are only within the purview of the few, there are other spaces in the city where used clothing is recycled in a more overt fashion *en masse*. The majority of the poor in India wear used, recycled clothing (plate 11) as the clothing charity Goonj has reported, and the weekend markets such as that at the Red Fort continue to provide for their needs. Second-hand markets are also found in Paharganj near the New Delhi railway station, on Qutub Road in south Delhi, and in the older parts of the city where recent immigrants have always settled, just as similar shops are found in London's East End. While some clothing is handed down by employers to domestic servants, many of the lowly paid maids, rickshaw drivers, and unskilled laborers in the city buy their recycled clothes from the old-clothes dealers here, Waghri men and women who barter for middle-class clothing in the suburbs and sell it on to the poor.

Located outside the Lal Qila (Red Fort) in Shahjahanabad, Old Delhi, was a long-established Sunday morning flea market, also known as the Chor Bazaar, or Thieves' Market. Until 2001, it covered several acres of waste ground at the back of the fort. Rubbish dealers, the *kabariwale,* spread their treasures out on the ground: a jumble of old watches, clocks, household utensils, old furniture, scrap metal, and books. Other dealers offered job lots of cheap china, umbrellas, and children's toys; one man habitually sold rolls of parcel tape that were surplus to someone's requirements, while another stall did a thriving business in complete sets of plastic airline crockery, with matching food trays, glasses, cutlery, and coffee cups from every carrier that flies into Delhi.

At the far end of the market a couple of hundred Waghri men and women sat on tarpaulins on the ground behind piles of washed, ironed, and neatly folded clothes. While sellers would call out to attract customers' attention, the area was noticeably quieter than the rest of the market. Indian retail clothes shops typically only display one or two garments at a time, with new clothes always folded and stacked on shelves inside. Customers must be tempted to stop and come in, and once they are sitting, the shopkeeper skillfully pulls out garments one by one with a practiced flick of the hand so the whole cloth, with all its decorative effects, can be glimpsed in the air as it falls in front of them. The folds in cloth are a signal of its newness, and the theatrical unfolding or unwrapping of a sari, a *salwar kamiz,* or a shawl heightens the performance by suggesting that the unfolding reveals cloth for the first time. Folds also denote cleanliness, as fold lines are carefully ironed back onto clothing after laundering. The Waghri pavement dealers use similar strategies, laying the folded used clothing out in front of them and tweaking out a particular size or color combination to convince a passer-by to stop, although piles become untidy by the end of a busy day.

Although a few daring middle-class bargain hunters might have visited the Chor Bazaar for amusement, they rarely bothered to venture to the end of the lot, where the Waghris sat. However, nearer the entrance there were also perhaps twenty to thirty stall holders selling imported used winter clothing, and many more tourists or middle-class Indians stopped here to buy. Stalls were piled high with sweaters, like a jumble sale, encouraging frenzied scrambles for the best buys. Some sellers hung trousers, anoraks, and fake fur coats up on hangers for all to see. The clothes had not been cleaned, but some were ironed. It was immediately clear that these clothes were not being marketed only to the absolutely poor who bought from the *patriwale;* they were being sold as fashionable Western garments to the aspirational lower and lower-middle classes. Traders and my acquaintances alike agreed that such customers could not afford to buy new the imported "real" brands that had become available in the past decade and of which they were increasingly aware through the media, but they wanted the quality and "authenticity" that were not matched by locally produced garments. If a pair of jeans from a top international brand, such as Levi's, retailed new for Rs 700 to 1000, a locally made, unbranded pair would be priced between Rs 150 to 300; the imported second-hand branded pair could be available for only Rs 50 to 70. Moreover, a carefully chosen second-hand garment could be passed off as one that had been bought new. No attempt had been made to remove the brightly colored paper thrift-store labels attached to most of the clothing. The foreign origin of the clothes was important to their marketing, although few of the people in the market would be able to read the labels—one trader enthusiastically sang out "Foreign!" every minute or two to attract buyers.

Many of the dealers were Tibetan and Nepalese; others were Muslims. The Muslims often also ran small shops in the Coat Market opposite the Red Fort, while the Tibetan and Nepalese traders used other routes to obtain such clothing. As imports are legal in all the countries bordering India but not in India itself, cross-border smuggling is common, and many of these dealers' families originate in the north.[16] In addition, many informants mentioned that charitable aid relief destined for Bangladesh was often diverted to India; one claimed that the whole illegal trade began on the borders of Bangladesh and Thailand. The largest volume of illegal clothing now comes directly via the port of Kandla in Gujarat, established as a Special Economic Zone (SEZ), and is traded through long networks of middlemen. Newer immigrants can clearly start off trading illicit goods on a small scale and make a successful business grow. One young Nepalese man had obtained a few old clothes from an older cousin, who ran a large stall in the market, and was selling them on commission a few yards down the row; others are paid a wage to run stalls spread out in a group to catch as many customers as possible.

Now, however, the Chor Bazaar has become a shadow of its former self, forced to move from the waste land since the terrorist attacks on the Red Fort. The Waghri now line the main road along the front of the

Red Fort, where they are harassed by the local police for illegal hawking. They now sit opposite the Coat Market, the more established second-hand clothing market, selling woolens for the winter and small lots to those peddlers traveling up to the hills. Tensions run high over accusations of bribery and corruption, since local officials apparently permit the Coat Market to operate with impunity. The contrast is striking to a casual observer: where the Waghri are squashed together on the pavement with little room to spread their wares, the men opposite construct high racks out of bamboo canes on which to display their coats and shoddy blankets made of recycled wool.

Wholesale Markets

Second-hand Western Imports: Azad Market

Used clothing is traded internationally as a commodity, originating in Western countries, such as the U.S., Canada, Europe, and Australia, and in Japan. Western dealers buy up unwanted clothing from charities or sale stock from manufacturers and sell it on, largely to the developing world, by the container load. Such companies are the direct descendants of the eighteenth- and nineteenth-century "rag and bone men," now trading in huge volumes in the global markets. The rag dealers selling imported clothing and the local Waghri both have their own wholesale markets, out of view of the public eye. Delhi's main business center for dealing in imported used Western clothing is located in Azad Market, on land behind Old Delhi Railway Station that was allotted to a group of traders by the government. This market was established after Independence largely for the auction of government surplus, and lines of shops still sell off uniforms, boots, equipment, tires, tents, shoddy blankets, and the ubiquitous red-and-blue-striped floor mats.

These clothes are prized; because they are of Western origin they are different, new, and potentially fashionable. Some governments, such as those of Kenya and India, treat them as a threat to indigenous industries, while others, such as those of Pakistan and Bangladesh, see them as a welcome addition to an impoverished economy unable to afford new clothing.[17] Many of the importers I spoke to in the market were sure that such clothing is needed by the poor, who cannot afford to buy new Indian clothing, which adds a positive moral dimension to their illegal trade. It seems there is never enough used Indian clothing in the market, and its quality is very low by the time it has been thrown out. The apparent craze for foreign things is explained by the dealers as primarily due to their cheapness, with fashion playing a role in only the more developed niche markets, and the reason for their popularity is the abject poverty of most of their customers.

Many importers are undoubtedly operating in the black market, since import taxes remain prohibitively high. It is legal to import mechanically slashed woolen clothing known as "mutilated hosiery," which provides wool as a raw material for the shoddy (remanufactured yarn) industry located in Haryana (Norris 2005a, 2005b). Under the cloak of this invisible industry, vast quantities of cast-off clothing are brought into the country. In a rabbit warren of cupboard offices in an old covered market, businessmen sit surrounded by phones and towers of plastic-covered bales of clothing, small boys keeping up a constant flow of chai. Smaller bales, neatly strapped, have their origins emblazoned across them: "CANN-AMM Best in Used American/Canadian clothing," complete with flags and national colors. These dealers use such bales as samples, maintaining huge *godowns* (warehouses) in the city outskirts. Networks extend across India (with distribution channeled through Kandla port) and through to the UK and the U.S.; many Indians living in the West are increasingly acting as agents in the old-clothing trade, and along the U.S. eastern seaboard many of the traditionally Jewish firms are now owned or operated by Indians.

It is clear that in this underground trade the importers and dealers are often outsiders, of diverse origins and traditions, and the trade appears to develop as new waves of economic migrants enter the business; it offers an opportune starting point for those with little to invest. There is a little cross-over with the dealers in used Indian clothing. The systems of collecting, processing, and importing second-hand clothing and waste products remain separate from those involved in the resale of Indian clothing or the export of products made from saris, but they do offer a route for upwardly mobile Waghris to invest in bales of imported clothing and realize greater profits.

One dealer had worked as an agent for imported used clothing for thirty years—his grandfather had started in 1955 after fleeing Pakistan following Partition, and his father had carried the trade on. Other traders recounted similar stories dating back to the founding of the market after Partition, and businesses were certainly extended family concerns in the traditional manner; older relatives started with nothing, haggled for government contracts, began importing emergency relief after Partition from the West, and had continued ever since, despite subsequent changes in the law. The notoriously complicated (and corrupt) former system of granting import and export licenses (the "permit Raj") has undergone major changes recently, following economic liberalization, but tracking the expansion of firms since the 1950s is complex—if dealers had no import licenses, they bought from those who did until they managed to acquire licenses themselves, and only one was willing to admit outright law-breaking. This informant claimed to build in to his costs the customs fines he regularly paid for misdeclaring his imports as rags, and his turnover was high enough to make the strategy viable.

Indian Cast-Offs: Raghubir Nagar

The final setting of my research was the suburb in which used Indian clothing was traded, containing a wholesale market, a large temple, and residential streets (figure 1). More than forty thousand Waghri people are directly involved in the business of collecting old clothing from Indian households in Delhi through *pheriya,* itinerant door-to-door trading, usually in the suburbs. They themselves live in a northwest suburb of Delhi, Raghubir Nagar, which is the major hub of old-clothes trading in north India. Thousands more families in nearby localities earn a livelihood through associated activities. While the poorest struggle to survive at the margins of the business, dealing in scraps day-to-day, more resourceful, established entrepreneurs transform unwanted clothing into new products for wealthier markets, constantly striving to increase their profit margins or cross over into dealing in imported goods.

The existence of a thriving Waghri community of old-clothes dealers, with their own market, temples, local council (*panchayat*), and support networks, is completely unknown to middle-class Delhiites, who treat the *bartanwale* (those who barter kitchen utensils for clothing) with undisguised disdain. Such lack of knowledge or interest, as well as disregard for the women sitting on the pavements, are key elements in the ability of the dealers to remove unwanted garments beyond the pale. Previous owners do not have to confront the eventual uses of their cast-offs, and for the large part believe that the items can travel only down the social hierarchy, providing cheap clothing for the poor and needy.

It is clear that there has always been a trade in old clothing, albeit more limited formerly than today. At the top end of the scale, a trade in rare and expensive textiles has long flourished, connected to aristocratic courts and the urban elite, with buyers from the West since the precolonial period and indigenous collectors post-Independence. There were also barter systems that allowed the poorer members of society to obtain something in exchange for old clothes and rags.

The Waghri are a Dalit caste, and are known throughout India as unskilled laborers and itinerant petty traders, some eking a living from the "rag and bone trade" or selling vegetables and local cigarettes, *beedis,* on the roadside. In their capacity as the removers of society's rubbish and filth they are often both despised and mistrusted. One prominent Waghri dealer suggested that old clothes had been bartered for steel utensils at least since the 1920s, when his great-grandfather had started out. He recognized that rich women would not just sell their old silk saris for a few rupees, but were willing to swap them for highly desirable metal pots, probably of brass and copper. Cast-off belongings are traded away from the locality, ensuring anonymity and precluding the chance of their former owner encountering them again, making the system doubly attractive. Tarlo's investigation into the beginning of the barter of pots for Gujarati embroidery in the 1950s shows how this particular exchange fortuitously built upon those preexisting systems of bartering unwanted goods, including clothing (Tarlo 1996b). Recent research in Mumbai has catalogued barter systems exchanging new goods for old, including biscuits for used plastic bags, garlic for other plastic items, and plastic plates for niche items such as old clocks, as well as more direct new-for-old part-exchanges (Gupta 2009).

Older members of the Waghri community involved in *pheriya* recalled how the trade has developed in Delhi since Partition and Independence. At that time, members of the Chamar caste used to sit on the pavement by the railway station, offering cups and glasses of chai in return for old clothes. It is uncertain how long that trade had been going on, but they implied that during the upheavals of the time, distress selling was widespread, and bartering was commonplace where cash was in short supply. As leatherworkers and processors of animal carcasses, the Chamars were traditionally perceived as untouchable in orthodox Hinduism, and it not unreasonable to assume that, as such, they were well placed to receive others' cast-offs, while the glass, china, and later steel utensils they offered in return could have been acceptable to those sensitive to ritual pollution, being made from materials conceptually less permeable to pollution than cloth. The Chamars would then sell the rags on to the *truckwale* for cleaning engines. The Waghris apparently took over the trade from the Chamars by continuing their traditional occupation of nomadic dealing and proactively taking their baskets of steel with them on *pheriya,* encouraging people to part with clothing for an attractive return. Often older men would comment that it was the resourcefulness of the Gujaratis, and in particular the strength and stamina of the Waghris, that ensured their success. Although *pheriya* is certainly a physically tiring trade, the Waghris are also subject to a great deal of suspicion and verbal abuse, which they have to face daily in order to survive. Most importantly, they have been able to maximize the value of different clothes through increasing the specialization of markets and diversifying trade to accumulate resources, creating new economic and social hierarchies within their community.

The Waghri community comprises distinct groups of sub-castes and extended families who have migrated to Delhi since Independence. During the many days I spent interviewing families in Raghubir Nagar to establish an initial outline of how the trade had developed, it became clear that the migration following Partition provided a significant stimulus for its expansion. The first influx of Waghri refugees came directly from the Punjab and Sindh provinces that had been incorporated into Pakistan, and they settled on the roadsides around the railway station and in the neighboring district of Paharganj. By the 1950s the traders appear to have moved out to *jhuggis* (slums) in districts encircling the modern city center: Karol Barg, Patel Nagar, Sadar Bazaar, and Ramnesh Nagar. During the 1960s the government made efforts to remove the slums in these areas, and land was allocated to the Waghris in Raghubir Nagar. At that time the area was still countryside, an "unsettled jungle," in the words of one informant. Each family was allocated a plot of twenty-five square yards on which to build a house.

During the decades post-Partition, more Waghris joined them from refugee camps in Jaipur, Alwar, and towns across Rajasthan. Refugees from the same camp often came from the same district in Pakistan. These groups would live in slums until a new allocation of land was made, continually enlarging the Raghubir Nagar settlement. The same groups of extended families still live together, tracing common migratory patterns and continuing to intermarry. All speak variations of Gujarati and were willing to identify themselves as such. Today, huge numbers of Waghri families have moved to the area from Ahmedabad and Gandhinagar in Gujarat. Many appear to regard living in Delhi as a temporary arrangement, maintaining joint family homes in Gujarat, perhaps leaving children or siblings behind, and returning for a few months every summer. As more and more rural migrants move to Gujarat's main cities, it becomes more difficult to find work, and the difficulty has been compounded by the large-scale closure of Ahmedabad's mills in the 1980s. Making a living from the clothing trade requires extremely hard work and the willingness to travel, and whole families are usually involved. The increasing number of traders leads to more competition and lower returns, but this is precariously balanced by both the growing surpluses of clothing in middle-class wardrobes to be teased out and the development of new markets for used cloth.

Alighting from the bus in the center of Raghubir Nagar, one is immediately struck by the multitude of clothing decorating the landscape (figure 7). Hundreds of pairs of faded blue jeans adorn the barriers down the middle of the main road, and rows of tiny shops selling steel utensils, plastic tubs, glass bowls, and china cups display their wares in serried ranks stretching yards out into the street, surveyed by their owners, who sit behind huge sets of pan scales in the rear. Along the narrow alleyways each small house appears to be wrapped in lengths of washing-line festooned with clothes (plates 4, 9, and 10). One sports imported old ski jackets, another local men's trousers, shirts, *salwar kamiz,* or school uniforms. Through open doorways one sees piles of clothing bundled up in knotted old saris, or rows of worn old shoes. Further into the dark houses, displays of shining steel utensils sit on shelves high up on the back wall, dazzling in the odd ray of sunlight.

Opposite the bus stop is a marketplace, which in the mornings is full of women sitting cross-legged on the ground in rows, selling the clothes they collected the day before (plate 1). The buyers are all men, trawling up and down, singing out what they are looking for, bargaining for pieces, trying to evade the women thrusting unwanted items into their paths. Along the crowded pavements outside, rows of men stand cleaning their teeth with neem twigs, drinking cups of tea, and hawking and spitting. Music blares out across the market from the temple loudspeakers every morning, and the name of each dealer who donates a small sum as he leaves is announced as well. With bundles of clothes in their arms and more draped over, the buyers eventually depart the frenzied market, clambering into rickshaws piled high with their purchases or trudging off down the road. This apparently chaotic scene is the heart of what is the largest indigenous used-clothing market in the subcontinent, attracting dealers from across north India. Drawing in unwanted clothing that has been bartered for across the country, the market reenergizes the latent value of cloth, filters and sorts it, and acts as a powerful centrifugal force, ejecting tens of thousands of items every day.

Each new group of Waghri settling in Raghubir Nagar moves into crowded squatters' colonies constructed from recycled materials, with earth floors and open sewers in the alleyways. The occupants have few resources and no storage space or facilities, and all are engaged in the daily *pheriya,* taking baskets of *bartan* (kitchen utensils) on rounds throughout the suburbs. Although known to middle-class housewives as *bartanwale,* within the Waghris they are the lowest group, the *pheriwale,* or "the ones who go on rounds." Government bureaucracy and corruption combine to ensure the fragility of their livelihoods. In order to apply for identity cards (and their attendant benefits) citizens require a ration card; to be eligible for a ration card one needs an address; but *jhuggis* are not generally recognized residences. After several years slums acquire a permanent character, shacks become houses and acquire numbers, and if residents are extremely lucky they may eventually qualify for an entitlement to land. However, recent plot allocations have been even further out from the city, away from Raghubir Nagar, and hence untenable for those in the business. The more established families do have legal plots of land, with enclosed sewers and brick houses complete with concrete floors and electricity. Their children go to school, and few of them engage directly in *pheriya.* These families invest labor and resources in adding value to second-hand clothing for resale, processes that I will return to later on. Some successful shopkeepers and traders have moved out of the used-clothing business altogether and just serve the community, but one local official informally estimated that 95 percent of the population deals in cloth.

The Waghri were given Raghubir Nagar, in which to establish their community, in 1962, during Nehru's premiership. In 1963 the Ghora

Mandir (Horse Temple), dedicated to Baba Ramdev, was begun opposite Raghubir Nagar; built of white marble and paid for by donations by prosperous locals, it was inaugurated in 1964. (Ramdev is a north Indian god popular with Dalit groups and often linked to Krishna.) It is run by a council, the *pradhan*, which also controls the local *panchayat,* settles disputes and collects fines, administers support networks, and negotiates with the municipal authorities. The business focus of the Waghri directly and indirectly centers on the daily market opposite the temple, the Ghora Mandi (Horse Market), where used clothing is bought and sold (figure 1). The trade had developed as the Waghri moved in, lining the roadsides and blocking the traffic daily. In 1975 the Delhi authorities allocated a garden area for trading, and more recently it was paved with concrete. Half the area was roofed, but the roofed part is used only during the rainy season, when it leaks profusely. *Chowkidars* at the gateways charge an entrance fee of two rupees to every dealer, which goes into a fund run by the *pradhan* that is intended to eventually buy the land from the Municipal Corporation of Delhi. The guards estimated that between one thousand and fifteen hundred dealers entered the market on any day, depending on religious festivals and the season, and this tallied with my own rough calculations.

Earning a living is undoubtedly exhausting for those *pheriwale* with the fewest resources. It is almost always the women who ply the trade, sometimes taking grown-up children with them, leaving others behind with older siblings. Whilst many men seem to work in associated businesses, they rarely go out on suburban *pheriya,* and drinking and violence are claimed to be widespread among the male traders. Some men deal in non-clothing goods, such as second-hand electrical goods, records, shoes, and watches, which are also obtained through the women's bartering, though this is a more recent trend. But unlike the usual trade in rubbish and scrap carried on by male *kabariwale,* this doorstep trade undoubtedly benefits from the fact that women are dealing with women, which increases the traders' chance of access to the domestic sphere and intimate possessions.

Every day the women get up long before dawn to take the previous day's clothing to the marketplace to sell. They begin arriving from 3:00 AM and stay until 9:00 or 10:00 AM, when they have sold them all. They then return home to the neighboring alleys, to cook a meal and prepare for the *pheriya,* often traveling via the *bartan* shops to negotiate for more stock. By mid-morning they clamber into rickshaws or buses, although they are often abused by bus drivers who refuse to allow them on with their overfilled baskets of *bartan* on their shoulders. They may spend up to two hours in Delhi's traffic, traversing the length and breadth of the city. Families clearly earmark patches of territory for themselves; on arrival at their patch, the women spread out their wares on the pavement, encouraging the local middle-class women to stop and inspect them. Many complained of being harassed by the police for having no license to trade, having to bribe them on some days, and they were often suspected of local robberies and kidnapping. By 9:00 or 10:00 PM in the evening they return to Raghubir Nagar with bundles of clothes and unsold *bartan,* to do the household chores.

The poorest families live hand to mouth, unable to amass clothing or *bartan,* with no access to running water to wash garments and no sewing machines or space to mend them, constantly begging every day for more credit at the *bartan* shop where they buy their supplies. There are more than one hundred and sixty steel dealers in the area, and some women run up accounts with several at a time. The better the quality of steel and plastic and the wider the range of utensils on offer, the more likely a suburban housewife will be to stop on the roadside, and the more likely it is that a garment dealer will obtain good-quality clothing in return. The *pheriwale* therefore aggressively scrutinize each piece in the *bartan* shop for dents, scratches, and ill-fitting lids before buying it. Steel is usually sold by weight, and the illiterate women had sharp memories, both for how much money they owed the steel dealers and for what each *bartan* was worth when they bargained for clothing.

Having described some of the main contexts in Delhi in which my research took place, I now shift my focus to the manner in which these sites are related and the flow of material through them. The manipulation of meaning in clothes is densely implicated in their fate by the women who own them, the dealers who trade in them, and the designers who re-create value through their reuse and recycling.

FIGURE 3. Bulbul's almari, a Godrej cupboard. PHOTO BY LUCY NORRIS

3 LOOKING THROUGH THE WARDROBE

Clothes' ability to project a desirable image is simply one element of their potential value. Any outfit or individual item of clothing can be sartorially useful for its smartness, crispness, and cleanliness; it may be fashionably new or respectfully traditional, suitably cool and comfy or stiff, shiny, and formal, depending on the context. But clothing has a range of subjective values associated with how it was acquired or where it was worn, conjuring up images in the mind of the wearer that can alter how she feels about a garment. Clothing is strongly associated with people in India and can come to both represent and substantially constitute relationships; as the embodiment of social ties and emotionally rich, the flow of clothing through the wardrobe and the choices of which things are kept and which handed on must be carefully managed.

There are invisible, intimate associations with clothing: the biographies of clothes reveal histories of swapping, lending, and borrowing, inheriting and handing down; clothes can act as souvenirs, mementoes, and mnemonics of past relationships and selves through their very materiality, their physical condition, color, smell, and texture (Hoskins 1998). Clothes that are actually worn are part of a larger accumulation of clothing in the home that includes rarely worn wedding saris, piles of unworn gifts, favorite clothes that are past their best, and unsuitable and unwanted garments. Individual items of clothing can have a sentimental value even if they can be no longer be worn; equally, there might be no particular emotional attachment to a sari that can be worn to work every day.[1]

Women's clothing fell into four main categories. The first consisted of saris, with their associated blouses and petticoats. Next were regional tailored styles such as the *lehenga choli*. Third was the three-piece stitched Punjabi suit, and European-style clothes such as maxi-dresses, skirts, jeans, cotton blouses, and T-shirts made up the fourth category. All my informants knew the exact history of each and every piece of clothing they owned, and could construct a narrative of their own lives through the circumstances in which their clothes were acquired.[2] When the doors of a wardrobe are opened, all past, present, and future clothes are synchronously presented as technologies of the self; the process of classifying them, of weighing up the value of each piece and according it its appropriate place, permits the successes and failures of the ontological project to be evaluated. Former selves, fantasy selves, rejected selves, and those immanent, with the potential of coming into being, are all displayed. "Objects hang before the eyes of the imagination, continuously re-presenting ourselves to ourselves, and telling the stories of our lives in ways which would be impossible otherwise" (Pearce 1992, 47).

If the point of the souvenir or memento is remembering, the point of the collection is forgetting (Stewart 1993, 152). In collections, each element is representative and works in combination toward the creation of a new whole that is the context of the collection itself. While the souvenir looks backward, the collection exists in the present by means of its principles of organization, defining space and time in terms of the collection's own internal dynamics (Benjamin 1992b); collections replace history with classification (Stewart 1993). Whether a garment is valued for its individual emotional associations or for its suitability as an element of a collection is important for understanding the means by which its value can be exchanged across registers and whether it should be kept, passed on, or got rid of completely.

Meghna's Wardrobe

Meghna was a married woman in her early fifties. She was the principal of a school in south Delhi, and every day she would pass me on the stairs on her way to work. A tall woman with striking features, she was always perfectly presented in an understated sari, starched cool cotton in summer or thicker, deeply colored silk in winter. After many attempts to set up an appointment, we found a morning during a school holiday to look through her clothing.

Meghna was born and brought up in Delhi, but had lived in Jaipur for a few years during one of her husband's military postings. She lived with her husband and Indira, her grown daughter, in the Progressive, although Indira had just finished a journalism course and was about to leave home for a job in Bombay. Her clothes were mainly kept in a large, built-in metal cupboard in her bedroom. Since I visited her in early February, she explained that she had all her winter clothing out for use, and most of her summer clothing was packed away in suitcases. She had also packed away most of her heaviest trousseau silks in a trunk to keep them safe from insects; most of them she hadn't worn since a couple of years after her marriage in the early 1970s, as she disliked the flashy, showy gold and silver *zari* work (embroidery with gold or silver thread) and the bright colors. She favored "lush" colors such as greens, maroons, and royal blues over "loud" colors such as reds and bright pinks. As a professional woman, she wore only saris to work and preferred them to suits, of which she had only about a dozen, which she wore mostly at home or for local shopping. Unlike some women who had moved to Delhi from different regions, she had no sets of Indian stitched clothing such as full skirts, blouses, and veils, and no Western clothing left, having handed it all on to her younger sisters years ago.

Meghna liked more traditional styles of sari from the south, which were also longer, better for draping and accommodating her height. She chose printed silks for work; she didn't like the heavier northern Banarasi silks, as they tend to be too ornamented with gold and silver. In particular she liked the printed *khadi* handspun silks, showing me one she had had for eight years that had been given to her by her husband, and another she had had for ten years. Although the good-quality silks cost more, she saw them as investments, as they last much longer. She was often given saris at festivals and on special occasions, and said that in addition she bought herself two or three a year, at which point her daughter interjected, "She buys them *all* the time!" Still, Meghna and her sisters were continually swapping their saris to give themselves more variety, although one, annoyingly, was not very good at looking after them.

She pulled out particular saris to show them to me. A plain one in a shade called "onion pink" had been given to her by her mother. Her mother had also passed on one made of synthetic crepe, which Meghna had had decorated with single-thread embroidery, but she had since got fed up with it. Two or three other silks she was tired of she had recently given to Indira, to have made up into suits by a friend who designs and stitches them in her workshop. She showed me one of her wedding saris, a south Indian silk with a "temple" border (triangle designs representing the pyramids, *gopuram,* found in south Indian temples). Very heavily worked in gold *zari,* it had cost about five hundred rupees twenty-six years before, and she thought it would probably cost seven or eight thousand if bought new in 2000. Although she would never wear it again, she kept cleaning and refolding it to keep it nicely. She also had a finer Banarasi silk from her engagement. She had received twenty-one heavy silk saris as a trousseau from her parents, and eleven heavy ones from her in-laws;[3] in addition she had been given many cotton and silk ones for daily wear. A particular favorite was a plain green one with a small gold border that she had worn at least thirty times since her marriage.

Meghna estimated that she had at least a hundred saris, probably more, and looked rather embarrassed at having to count them. When we discussed summer and winter clothes, it became clear that she had easily more than a hundred saris for each season in her everyday wardrobe, with her wedding saris and party clothes in addition. A few more were torn or faded through daily use and kept to one side; she liked to "keep a collection of new saris going so that they don't all cave in at once."

Maintenance and Care of Clothing

Clothes bear the imprint of the day; they collect the creases, smells, dirt, and dust of metropolitan life, and the blood, sweat, and tears of the body. The maintenance of clothing to present oneself as well as possible is a labor of love, to which women devote much care and attention. Shakuntala, another of my informants, was a young woman who wore saris every day. She was an NGO officer and media consultant. She was a Rajput from Uttar Pradesh who had made a "love marriage" to a Bengali against his parents' wishes two years earlier.[4] When she first married, just putting one on took her so long that she resented the time and effort involved in their care. Maintaining them can also become expensive: when I got to know her she lived on a tight budget in a very small two-room flat in south Delhi and washed her saris herself, then took them to the dry cleaner for starching and ironing at a cost of Rs 10 each. She had a system, wearing each one twice to work and taking eight every fortnight to the cleaners. When ironed, saris are folded up and stored flat on a shelf or across the bar of a hanger. Before wearing they may need yet another light iron before she goes to work.

Washing cotton saris and suits is time-consuming and arduous, especially in summer, when most clothes are washed after each wearing. Many women in the Progressive employ maids to do their daily washing, which is done in the bathroom with cold water, blocks of soap, scrubbing brushes, and wooden beaters, *danda,* to pound out the dirt. Such treatment itself wears out cloth, especially the looser handloom cotton favored by the upper middle classes, and washing machines and detergents are increasingly popular. Clothes are then hung out to dry on every available balcony or line before pressing.

Cotton saris need starching, which is sometimes still done at home using rice water or commercial starches, but may also be done by local cleaners who starch them and stretch them out on wooden batons and pegs before rolling them tightly up to dry before ironing. Silk saris and suits must be dry-cleaned. As this is both costly and thought to shorten the clothes' life, greater care is taken to treat them well and avoid cleaning them too often. They too are folded and kept flat. But in addition to being kept clean for wear, many silk saris, such as wedding saris, are rarely worn; the care taken of them is intended to prolong their life. Usually they are folded together and wrapped in protective cloths made from soft old cotton *dhotis,* saris, or muslin squares, occasionally stitched into bags. Silk must be protected from insects, and repellents include slivers of soap, mothballs, and leaves from the neem tree.

There is an annual cycle of care that includes refolding saris to prevent the silk, which tends to become more brittle with age, from tearing along the stiff creases. Around the time of the festival of Diwali, at the beginning of winter, silks are unpacked and aired as part of the changeover in seasonal wardrobes and general cleaning of the house for the celebrations; then they are carefully folded and wrapped up to be put away for the summer some time around Holi, the festival of color. Some women refold clothes more frequently. Saris with silver threads should be sent to be polished, as the threads quickly become tarnished and damage the fabric, but nobody I spoke to ever admitted to actually doing this; it was part of the lore of care that was unachievable in practical terms, and in fact few saris contain real silver and gold *zari* thread anymore.

The proper care of silk saris is a responsibility felt by nearly all women; though many women take pride in doing it well, others admit with some shame that they do not adequately look after their silks and feel that they should do better. Women recalled the time their mothers spent running the household, washing, drying, and folding the whole family's clothes, and good housewifery skills are a matter of pride. The repetitive daily, weekly, or seasonal attention paid to clothing enables them to be closely inspected for damage and any necessary repairs made. Each piece of clothing in the family is well known to a woman, and it is her investment in it that preserves it; the acts of routine maintenance help to determine which pieces require extra effort to be further treasured and which may be consigned to a lower category or more speedily discarded.

Wardrobes: The *Almari*

Indian women keep their clothes and precious items in a wide variety of containers: large wooden chests, pots with sealed lids, stacks of tin trunks, or built-in cupboards. The typical contemporary urban wardrobe is a lockable cupboard called an *almari*, often made of metal;[5] Godrej is a popular brand (figure 3). The ubiquitous grey, green, or brown steel cabinets bear a strong resemblance to a large office safe, having thick double doors, and at approximately seven feet high, they dominate bedrooms in apartments, which often have a special niche built to house one. Inside they have various combinations of a shallow hanging rail, lockable metal drawers, and shelves. A Godrej forms the core of a woman's storage system, and most of my informants owned and used one.[6]

The Godrej cupboard is itself a costly item that often forms part of the bride's dowry.[7] It is a movable entity that symbolically marks the beginning of a couple's life together and bears testimony to the passing of time, the arrival and departure of children, and the continuity of the family.[8] Bulbul, an Assamese woman, had received a Godrej cupboard from her parents on her marriage. Immediately afterward she moved to Delhi with her new husband. The cupboard, the key to it, and the clothing it contained, together with all her dowry of furniture, rugs, and utensils, remained with her mother-in-law in Assam. In the first few months of her marriage, she had nowhere to keep her trousseau of new clothes and jewelry sets, as she had no Godrej. She said she didn't really feel like a newly married woman until the cupboard could be delivered by rail, arriving bearing the scars of the long journey across the subcontinent.

The cupboard's contents, clothing, jewelry, and money, are valuable possessions. One of the functions of the cupboard is to protect its contents from outsiders, an ambiguous concept when a daughter-in-law moves into her new husband's home. Every woman to whom I spoke kept her cupboard locked, explaining that one could not trust servants in the house not to steal things. Where extended families live together, the habit also protects one's possessions from unwanted prying and borrowing by female in-laws. But where clothing is perceived as a common resource, a bride may have to hand over the key. Wearing a daughter-in-law's trousseau is definitely considered shameful amongst the families I spoke to, but it is acknowledged as sometimes necessary. Twice I was told of incidents where "other women" had had to quietly allow their mothers-in-law to use some of their unworn wedding saris as gifts to a new sister-in-law, which can cause some resentment and bitterness.

During the months while Bulbul was waiting in Delhi for her cupboard to arrive, she knew that her mother-in-law had been wearing some of her wedding trousseau, both saris and sets of the Assamese two-piece *mekhlar chaddar*,[9] that she had had to leave behind (see also Tarlo 1996a, 180). Her in-laws were an artistic family, and they had specially commissioned many of the counter-gifts from skilled weavers to welcome their new daughter-in-law. Indeed, Bulbul suspected the delay in shipping the cupboard was prolonged by her mother-in-law, who wanted to keep many of her wedding gifts in her own home; although Bulbul took home as much clothing as she could after each visit, much of the furniture from her dowry was never sent on to her. When she later divorced, her father had to threaten legal action to get the wedding goods returned.[10] She now lived in the Progressive with a young child and a maid; she had trained to use a computer and did small secretarial tasks for her neighbors.

In the *almari*, saris are folded over hangers or wrapped in bundles of old cotton cloth and blouses and petticoats are arranged in piles, as are suits with their matching scarves. Tough Godrej suitcases are also popular for storing valuable clothing worn less often, and are kept under beds, high up on shelves, or even inside locked built-in cupboards, a double containment that emphasizes their value. Women such as Shakuntala, who had fewer garments and had not been married very long, would use the Godrej as a main wardrobe for all their clothes. But the amount of clothing women own increases with their age, and older women would start to divide it up into smaller sets, archiving clothing less often used or out of season in cases, trunks, chests of drawers, and cupboards. Frequently women revealed cases shoved under the bed, stuffed with saris, hoards of cloth that had been received but never worn.

Bulbul's storage of her clothes was unusual but revealing, as she utilized various rooms, niches, and pieces of furniture to manage her various personas, acquired through relationships with her family, ex-husband, and boyfriend. Her Godrej cupboard contained her expensive, formal wedding trousseau pieces, jewelry and silk saris given to her by her parents and in-laws. Once she was divorced, it represented a different self, pertaining to the married woman she no longer was. She had moved it into her second bedroom, where her child played and guests slept on a mattress. The clothes she actually wore were largely suits and Western clothing bought for her by her boyfriend, and she kept them in her bedroom, in a chest of drawers and in nooks and crannies hidden by curtains and old saris.[11]

Accumulating Clothing

The management of clothing and the wardrobe is clearly a technique for asserting both the individuality of the self and the importance of social relations, but in what way can clothes be considered relational objects? How wardrobes grow over time, together with how clothes form links among those who give and receive them, changes the perception of each piece and its status vis-à-vis other garments. Clothing is acquired in a variety of ways, a continuum marked at one end by formal (social) gifting at recognized stages in a person's life and at the other by informal, individual, and everyday transactions.[12] Janaki, a college teacher in her late thirties who lived with her young son Arun in the Progressive, described how her wardrobe grew:

I grew up with a limited wardrobe. I had perhaps four sets of
suits and my school uniform. It was not due to the lack of money,
but it was just that more was not necessary. At high school and
during my undergraduate years, I had two pairs of jeans, five or
six shirts, and about three suits. I wore a lot of my cousin's hand-
me-downs, and my brother wore Dad's clothing. I never had new
clothes. My first two saris were gifts; they were old ones of my
mother's. When I was doing my master's degree I had a scholar-
ship, and began buying my own clothes. My wardrobe grew, not
by buying from boutiques, but from Janpath stalls. [The market
for cheap, "export surplus" Western clothing is off Janpath, in
the center of New Delhi.] I was wearing Western clothes—I was
always a tomboy, it was a big part of my image . . . still is. In 1987
I got my first job, so my wardrobe expanded: I had to wear saris
to work, a sartorial revolution. I had to make an impression—if I
wore a suit I got mistaken for a student. But it was confusing, as I
was conferred adult status through my job yet I was not married.
On my marriage I acquired twenty-one saris from my parents
and a few from my in-laws. . . . [Through gifting and shopping for
myself] I now have over 150 saris.

Although the second half of Janaki's account focuses on the clothing
she has bought herself as a working woman, in fact a tally of her "summer
wardrobe" at that time revealed that of 72 saris, 39 had been gifted to her
by her mother, relatives, and close friends; 15 were part of her trousseau
from her mother; 8 were handed down from her mother; 9 she bought herself; and one was a gift from her former in-laws. When these numbers are
combined with the corresponding ones for her winter clothing, it becomes
clear that a significant majority of her clothes were acquired through gifting within her close family circle.[13] Once this had been balanced by counter-gifts of saris from her husband and his family, but since she was now
divorced, she retained only a single residual item from a larger number of
such gifts she had received upon her marriage. Janaki had been brought
up in Delhi and had married "out of caste" to a Bengali. At her wedding
she was given several Bengali saris that she would wear to her in-laws'
family occasions. They felt very different from what she used to wear, but
she liked them; she explained that these saris are renowned for their good

quality. When she got divorced her in-laws demanded all their gifts back, and as she no longer wished to have them in her house she was only too pleased to return them. Her divorce was acrimonious, and she wanted neither to dress as a Bengali nor to be reminded of her husband. Instead, she was very proud of her independence and increasing ability to purchase her own good-quality clothes, and took pleasure in saris her mother handed on to her.

Children tend to wear a mixture of hand-me-downs from elder siblings, cousins, and close relatives, who occasionally may give them new clothing as well, as presents from trips abroad or on family visits.[14] As the teenage years progress, they are gifted clothes by relatives, or they may go shopping with their parents or sisters for cheap jeans, skirts, and blouses in export surplus markets, but these are also passed on to younger children. As girls start to wear suits for some family occasions, they will be given or lent one by an older sister or cousin, and usually the first saris worn by a young woman belong to her mother.

It is at marriage that a woman begins to assume her full social status and needs the trousseau provided by her parents, counter-gifts from her in-laws, and presents from family members in order to fulfill her new roles and responsibilities.[15] Her parents may give her eleven, twenty-one, or thirty-one saris—or even more—that are expensive, of fine quality and heavy with gold and silver thread. These may have been collected by her mother over a period of years before the marriage, a sort of "wardrobe in waiting" added to whenever advantageous or unusual purchases could be made. Fathers may also participate in collecting for their daughters. Usha was in her late forties and taught management at Jawaharlal Nehru University; she and her husband both had large extended families in Bengal. Her father, who traveled the country as a doctor working for the World Health Organization, accumulated pieces for her trousseau over the years. From the time she began to think about marriage, at about twenty-two or twenty-three years old, he would bring traditional items back for her from the places where they were made.

After marriage, one of the major ways of increasing the wardrobe is by means of the gifts accruing to the multiplicity of roles a woman assumes throughout her life: new bride and wife, daughter-in-law, mother, mother-in-law, and eventually grandmother. Women receive clothes for all the major *samskaras* or life-cycle rituals of their children, from birth to marriage. If a woman is widowed, she may also receive gifts of new clothing that marks her new status, in white, cream, and, nowadays, subdued pastel shades, and no longer wear the bright shades appropriate for a wife.

Annual religious festivals such as Durga Puja, Diwali, Eid, Holi, Teej, Rakshabandh, and Janamashtami provide occasions for making offerings to temples and shrines and giving gifts to relatives according to regional and caste traditions. In north India, the head of the family is traditionally expected to give the whole family new clothes at Diwali, and a husband often gives his wife a sari or a suit for their wedding anniversary. On Rakshabandh, brothers gift their sisters with clothing as a sign of support and of acknowledgment of the brothers' responsibility for their welfare. The garments given in this manner are added to the wardrobe and can add to a feeling of comfort and support.

Saris are also appropriate gifts for women in professional contexts. Shakuntala, an NGO official, showed me one in her wardrobe given to her by her boss, who had distributed them throughout the project team after a particularly successful campaign. Government officers and politicians may give scarves and saris to distinguished visitors, friends, colleagues, and subordinates, usually men, who later pass them on to their wives and daughters, the garment standing symbolically for the status and success of the husband.

Family visits, especially annual visits or those made after long absences, are also occasions for the hosts to present clothes to their guests, usually at the time of leaving. This is particularly important when women return to their natal homes or meet their own kin, and the link from mother to daughter remains a major conduit for the acquisition of saris, new and old.[16] Get-togethers at festivals and in daily life provide opportunities for the constant swapping and handing on of clothing between sisters, mothers and daughters, mothers-in-law and their daughters-in law, aunts, nieces, and cousins.

A woman's core wardrobe of silk and cotton saris is gifted to her, and her style is thus defined by the nature of her relations with others; this basic dependency of women may require a combination of skill, subtlety, and assertiveness to overcome. In order to develop a wardrobe and be able to better articulate herself through clothing, she must have the appropriate social resources. Economic wealth is important, including women's ability to independently purchase their own clothing, but a wealth of kin is also essential: relations who will formally gift women with clothing. Madhuri provided a stark reminder of the consequences of the lack of family: "My mother's parents died while she was young, so she only had a brother [to provide for her]. Hence as a young woman she had only two saris and used to alternate them, washing them at night. When she got married to my father, he started to give her many saris, so in the end my mother had quite a large wardrobe."

Women may lack a core wardrobe of silks if they do not receive a trousseau because their parents do not approve of their marriage. Shakuntala's wardrobe contained only everyday saris and one suit. A photograph of her on a trip to the Taj Mahal in Agra with a college friend before her engagement showed her in jeans and jacket, her hair cut short in a smart bob, sneakers on her feet. A month or two later she had married, given away her whole wardrobe to her younger sisters, and completely assumed the dress of an orthodox Hindu wife. She now wore her hair long and braided and with the red line of the *sindoor* in her part, red glass and white ivory bangles, a *bindu* on her forehead, toe-rings, and anklets. She explained that her husband really liked her to dress this way, and she was only too happy to do so: she had married for love; the couple lived alone, ostracized by her in-laws; and she was proud to look and feel a married woman.

Her husband's family had given the couple nothing, did not attend the ceremony, and refused to see her now. Her own mother was willing to provide a trousseau, but Shakuntala felt it was unnecessary in the circumstances; it would put her to significant expense and she would not need to show off at family occasions to which she was not invited. She only accepted one wedding sari, so her wardrobe contained few expensive items. Instead, her husband had bought her everyday cotton and synthetic saris for every Durga Puja, Diwali, anniversary, and birthday since their marriage. This augmented her wardrobe and allowed her a greater choice of clothing, but one confined by her husband's taste and pocket. Further, since she had given away all her Western clothes and Punjabi suits at her marriage, she had had little to wear for several years.

Urvashi, a Hindu journalist from Lucknow, had married her Muslim husband secretly in a civil ceremony, without the consent of their families. She had no wedding sari and no trousseau, and had only recently been accepted into her husband's family and begun to receive gifts from her sisters-in-law. She had not had children and had never received clothes on any of the usual occasions, either religious festivals (neither she nor her husband had converted to the other's religion, and they had agreed not to celebrate such festivals) or motherhood celebrations. Like many women, she bought herself modest everyday suits and saris, but lacked a core wardrobe of good-quality silk clothing. After moving to Delhi and the Progressive, she was no longer working but training as an astrologist.

Women can acquire large numbers of clothes, depending on their circumstances. My most detailed wardrobe surveys make clear that a "working" wardrobe of at least 150–200 saris at any one time is quite normal for these women, including silks and winter wear, everyday saris, and older cottons. In addition, most had thirty to fifty suits and some Western-style garments; some women appeared to have many more than this available to be worn, but were unwilling to quantify them. The total quantity of clothing is greater once one includes those hoarded in suitcases and trunks that are rarely, if ever, worn—often another fifty to a hundred pieces, sometimes more. The women in the Progressive were largely from families possessing high symbolic capital in the form of education and professional capability, but were not particularly rich compared to families in south Delhi or wealthier suburbs. They kept good-quality clothing and made it last, and acquired more in order to be properly attired at functions, but they still had large numbers of surplus saris.

Managing Clothing

Wedding Saris: Treasured Bundles

A woman's best silks are usually known as her "wedding saris"; a young woman is expected to use her trousseau of wedding saris for formal occasions, and that is one of their main functions. She is the vehicle of her relatives' vicarious consumption (Veblen 1994). But as she grows older and receives new silk saris, the wedding saris may be relegated to semi-permanent storage, as may subsequent gifts of high-quality silk saris. Also included in this special group might be a sub-category of silk saris inherited from mothers and grandmothers, which may never be worn because of their poor condition but are, nevertheless, incorporated into the stock.

These bundles are significant in that they include saris from earlier generations as well as those that represent formative moments in a woman's own life cycle. Some of these silks may already be forming a new category in a mother's mind: those she is keeping for her daughter or future daughter-in-law. They appear to form an inalienable group of possessions closely bound up with a person's identity and status across time, and that is how they are talked about and the context in which they are shown (Weiner 1992). Yet as they become stained or torn, slowly disintegrating over time, memories of their previous owners fade and their inclusion in the category becomes more problematic. They may be put aside for getting rid of.

Despite the quantities of silk saris folded in trunks that should be suitable, women often claim that they need new ones. Ritually, new clothing is required for all major ceremonial events; it also keeps up family appearances. Uma, a Tamilian economics graduate in her mid-thirties who lived in the Progressive with her husband and their two young children, recently attended two family weddings held back to back, which lasted for two weeks. Every day the women wore silk saris, sometimes changing several times during the day, and no one wore the same sari twice. Although many women assert that "a sari never dates" when showing their clothes in the wardrobe, it can usually be dated. This ambiguity highlights the tensions inherent in women's sartorial decision making. One must wear new clothing to keep up appearances, whilst extolling the value of existing garments for their quality, aesthetics, and rarity. New saris offer endless opportunities for comments by others, whilst older saris, which display unusual or forgotten designs and combinations, may also be a talking point. Women's memories of others' clothing are acute; details of their designs, colors, and costs and the occasions on which they were worn are recalled and compared. A comment on a remembered sari might be a compliment, an acknowledgment of the woman's good taste or good fortune, but it might well also point up the woman's inability to purchase new clothing, to meet social expectations.[17]

Claiming that "the fashions found in older saris may come back again," Usha keeps them in a trunk, like most women. However, she admitted that by the time the fashions cycled around again, she would probably have acquired new silk saris and would not wear the old ones. Feeling that one has "enough" is extremely important, even though new acquisitions are always required to look one's best. Saris are the exemplary form of gifts for women, and replenishing the wardrobe through new acquisitions displays the extent of a woman's social networks, the support of her extended kin, and her own purchasing power.

Many women disliked wearing overly showy wedding saris, but felt sentimental about keeping them; only one admitted she had actually sold one unworn. Sangeeta was a teacher in a university college who had studied and worked in England. She was from north India and had married a Bengali whom she met abroad; they had two teenage children and lived in the Progressive. She had never wanted heavy wedding saris, but although they were generally "left-minded people," her husbands' parents had wanted to give her some; her kindly father-in-law had gifted her a very bright, rich, red and yellow Banarasi sari covered in *zari* work which she described as "horrendous." She had taken it with her to England in the late 1970s and sold it to one of the working-class Indian families she worked with for £25; she claimed that no upper-middle-class Indian would have worn it, and as Indian students they were very hard up in the UK.

Redundant Saris: Too Many Gifts?

Looking through piles of saris, I was struck by how many of them were never worn at all, or worn only once or twice before being packed away, never to be worn again. Women save the expense of getting a fall sewn in or a matching blouse made up until they intend to wear a sari, and often admit that some gifts are unlikely ever to be worn. Others are worn infrequently, perhaps not really liked; matching blouses become too tight and may not be replaced, making their wearing even less likely. There are many reasons that clothing in the wardrobe may go unused. Younger women may prefer suits and reject the sari outright in daily life, middle-aged women usually have simply acquired too many and can choose those they really like or are obliged to wear, and older women or widows may consider pale shades more appropriate and feel their former wardrobes are unsuitable. In addition there are changes in fashion, disastrous mistakes (whether purchases or gifts), and reminders of people one would rather forget. Specific saris may also have gone out of fashion, or have been inappropriate gifts at odds with the recipient's self-perception.

Many middle-aged women believe that saris are not hard to wear once you are used to them; Sangeeta related how her own mother used to play tennis in a sari. The older generation of women still sleep in saris, and claim that they make breast-feeding easier, as you can undo the *choli* and wrap the *pallu,* the decorative end, around your shoulders to keep warm. But despite claims of the sari's versatility, many young women feel that it imposes physical and social restrictions on them: it is laborious to maintain, requires effort to present a neat and tidy front, and can be awkward and cumbersome in a busy metropolitan environment. The sentiment that wearing a sari imparts grace and dignity, and that the outfit can be simultaneously attractive and authoritative, is not necessarily convincing for women used to Western clothes (which they may feel more free in) as college girls and suits as working women. For some younger married women, the art of wearing the sari with ease takes time to acquire, and conflicts with the desire to return to wearing suits. The birth of children is often the socially accepted time to do so, ostensibly to avoid the dribbles and spills of babies and the sticky-fingered tugging of toddlers.

The material that clothing is made of also affects its value, and the upper middle classes always favor natural fibers above all others. Silk is the most prized material (with real gold and silver as decorative elements), followed by cottons, terry-cot (a cotton-synthetic blend), and lastly polyester. In winter, a similar scale ranks *shahtoosh* (an extremely fine antelope wool),[18] pashmina, wool, wool-mix, and synthetic shawls and cardigans. Discussions about problematic garments frequently revealed that women did not know their exact composition, and new mixed-fiber fabrics appeared to be accommodated within the hierarchy according to their similarity to natural materials in such properties as shine and texture. For example, a cotton-rayon fabric with a sheen was described as "silky" and useful for a smart suit in daily wear, more stylish than an average cotton but not as luxurious as real silk.

Polyester saris have become a problematic category, and their rise and fall follows typical sumptuary patterns of elitism and availability. Usha explained that during the 1950s, cotton was cheap and plentiful, so the rich were not able to show off by wearing cottons. Polyester was a new fabric and very expensive in India, so it quickly became an elite status symbol. The yarn and technology had to be imported, but it was also costly because of the protectionist policies that supported cotton through subsidies and taxes on imports. Once the manufacture of synthetics expanded, government policies changed, and synthetics suddenly dropped in price. Polyester was then taken up by the poorer classes as cheap and easy to maintain, while most of the elite stopped wearing it as soon as they realized it was no longer a luxury. Now that extended families with servants have become rarer, handwoven cottons have become more of a status symbol because of their relative expense, laborious maintenance, and laundry costs, as well as their shorter lifespan.

Many women in the Progressive had synthetic saris in their wardrobes, usually received as gifts from "outside," that is, abroad. Artificial silks may be liked for their ease and soft draping, and are good for party wear; coarser polyesters may be rather apologetically produced as example of someone else's taste or as things bought some time ago from the Middle East. Women often commented that polyester is "not very nice" in the Indian climate, but useful during the monsoon period, when streets are running rivers, everything gets filthy, and clothing never dries.

Usha identified two groups in the Indian elite: those who eschewed synthetics (e.g., polyester) and considered them to be useful only as "wash'n'wear," and those who continued to wear the latest imported synthetics (such as artificial silks and expensive Japanese chiffons) for their sophistication and draping qualities. Not quite the highest Indian elite, but from solid upper-middle-class backgrounds, the women in the Progressive were caught between opposing value systems, some choosing to favor certain expensive synthetics as modern once again, others rejecting all synthetics as useful but unfashionable; there was clearly some confusion over their status. The functions that people most often attended were religious and family affairs, where silk saris are required; some women had one or two chiffons for wearing to a party, but unless one goes out a lot to parties, restaurants, and so on, they are not a worthwhile investment for a thrifty family, and have been historically very expensive in India. Similarly, although some silk suits available in boutiques may be made from extremely expensive fabrics and beautifully designed, many women, including Meghna, thought them to be useless and a waste of money. Cotton saris are the usual daily wear, and silk saris are expected at any formal gathering; expensive saris are gifted to women for this purpose.

Sometimes relationships may be affectionate, but tensions build up, and someone else's perspective on a woman and her taste is at odds with her self-perception. Rohini, a woman in her mid-thirties who lived in a double apartment with her husband, their two children, and her husband's parents in a nearby housing society, rather disapproved of her younger sister, conjuring up a rather stereotypical vision of the Punjabi culture she married into and her apparent concern to show off her new-found wealth. Her sister continually gave Rohini dun-colored suit lengths of synthetic nylon, which Rohini cannot abide. She wanted to recycle them by gifting the material to someone else, but her husband hated her doing it. She countered that she did not have enough room to store the cloth and would never use it. Certainly Rohini felt she could not wear the material outside the house and eventually asked her sister to stop presenting her with it, a difficult request. Rohini was uncertain whether her sister actually liked these synthetics, seeing them as rather modern and sophisticated, perhaps exotic; Rohini herself considered them cheap and lower-middle-class, and wondered whether the gifts were an attempt at a subtle put-down which backfired.

As women age, they are expected to move from married woman and mother to matriarch and grandmother. For many women, this is a period when they assume control over an extended household and are arranging marriages for their children. These women usually appear well turned out, often wearing more traditional saris and suits, carefully maintained and formally arranged. As figures expand and personal authority increases, clothing is less likely to draw attention to the body itself—for example, with clinging chiffons—and more likely to construct contours in space with well starched cottons and crisp silks. Older women often cease to wear bright saris and suits in daily wear, and overall they wear less embroidery and less metal thread such as gold and silver *zari*; opinions about the appropriateness of brightness and color for widows' clothing varied amongst my informants, but all expected widows to present a more muted appearance, with pastel shades and little ornamentation. These changes mean that more clothing is no longer worn and may be put aside.

Cosmopolitan Style or Regional Symbol?

Women's knowledge of their own regional sartorial customs is respected, and reflects alternative regimes of value in Delhi's cosmopolitan lifestyle. Almost all the women I spoke to had a category of clothing in their wardrobes they considered to be particular to their community of origin, either embroidered woolen shawls and Kashmiri tunics (*phiren*) or styles such as distinctively stitched skirts and blouses, and were eager to show them off as significantly different from the standard heavy Banarasi silk sari. Designs, fabrics, and colors associated with women's places of origin are often abundant in their wardrobes, and are occasionally worn as badges of identity. For example, Bengali women are famed for their beautifully woven cottons, while Rajasthani and Gujarati women are expected to have a preponderance of tie-dyed *bandhini* saris or block-print Kota checks. Susan had a small collection of white saris and *mundu* sets[19] from Kerala, where she and her husband were both from, elegantly decorated with gold borders and thick gold stripes across the decorative end, the *pallu*. Regional clothing provides a material link to home and family, especially as most are acquired through gifting. As the sari (now together with the suit) is the garment of choice for women in Delhi and throughout urban India, these variations usually remain unworn after the marriage ceremony, or become the favorite party attire of young girls and teenagers as yet too young to wear a sari. Urban sophistication has rendered them obsolete, and most women tend to wear heavy saris to functions after they are married.

Her trousseau pieces reflect a woman's own region, but most of the styles and patterns of Indian textile traditions are also represented by saris or suits in her wardrobe. Women living in the Progressive took pride in their extensive networks, which allowed them to acquire new and unusual clothing and demonstrate personal flair by incorporating such items into their wardrobes. The more far-flung and diverse the places a woman obtains her saris from, the more cosmopolitan she appears.[20] The wardrobe becomes a transcendent entity that represents a cultural kaleidoscope conveying a sense of "unity in diversity."

Saris and Western clothes are bought in shops and at fairs, though wealthier families might commission wedding saris from weavers or their agents. Suits are largely made to measure by tailors from cloth and decorative elements supplied by the customer. These can be bought as a predesigned set, or the women themselves can put fabrics and trimmings together. Many regional styles of garments and decorative schemes are available in Delhi in shops, boutiques, and the government and state emporia. Regional craft fairs, usually run by NGOs such as the Blind School, and charity shops like Dastkar provide both the opportunity to purchase new handmade clothes more cheaply and the satisfaction of having contributed to a good cause. Often women shop at these places in groups, and purchase clothes for themselves and friends. Unlike the south Delhi boutiques (such as those in Hauz Khas studied by Tarlo [1996a]) that mix and reinvent regional forms each season to create versions of "ethnic chic," the craft fairs and charity shops sell clothing that the women feel is more authentic. Moreover, they buy their stock almost directly from the artisan producer, via a local charity or government agency that is expected to be trustworthy and pay the producers fairly.

Many women want most of all to obtain regional styles from the regions themselves, either directly from the producers or from local retailers. Direct acquisition lends greater authenticity to the clothing and, of course, saves money by cutting out the middleman. Buying saris and lengths of cloth that will be needed for upcoming events can be expensive, and women try to find as many means as possible of reducing costs. They are extremely proud of saving money and constantly compare prices, both to share sources and to compete for status as bargain-hunters. So when friends or relatives make a family or business trip, the woman will plead for saris. In some cases, a traveler herself becomes a middleman, bringing back a suitcase full of bargains and inviting her friends around to purchase them. Radha, for instance, was a Gujarati woman in her early fifties, married to a Gujarati who was an academic; they had one young daughter. (She had been widowed much younger; this was her second marriage.) She frequently went back to Gujarat and brought home cloth to order. Sangeeta and others would tell her that they needed three or four saris in a certain price range, and Radha would select them herself. She would make a little profit, and her neighbors would shave a few rupees off the cost of each one.

However, Rekha, a wealthy young Punjabi woman who was married to a financial advisor and living in a large detached house in south Delhi with her in-laws, explained that when she went with her mother to choose her trousseau, she did not know much about textiles and patterns or what she liked. They decided to buy an assortment of different regional styles in a few color combinations—they had a "couple of everything" sort of attitude, she said, itemizing her trousseau as "one *ikat,* one *patola,* two Orissa, three Kanchipuram, two *bandhini,*" and so on. The choice is endless and can lead to a desire to have "a sari from every region in India." Of course this desire, familiar to any collector, can never be fulfilled, but it adds direction, drive, and longing to the continual process of acquisition (Benjamin 1992b). Rekha was a school friend of one of my neighbors, but came from a different social and economic background, and her attitudes to the value of clothing provide a contrast to my other informants.

Saris and accessories gifted from abroad are even more alluring, because they come from such a distance. The Middle East, particularly Dubai, Riyadh, and similar oases of duty-free consumption, provides a steady stream of synthetic saris in pastel shades, brought back by relatives on stopovers or who have relocated on business. The more recent imports from Japan and now China are similarly valued, especially as they may be significantly cheaper.

Knowledge of Indian regional textile techniques and designs is often a sign of belonging to an elite social class whose sophistication and discernment arises from growing up in a family that traveled and habitually purchased good-quality clothing. The desire for novelty, authenticity, and good value can be taken to the extreme, until the saris in a wardrobe become a nascent collection, full of gaps, and conceived of as only a tantalizingly partial sampling of all that is available. As regionalism remains important, so nationalism counterbalances it, and it remains a matter of pride to incorporate foreign saris into one's wardrobe. These interlocking layers of geographical identity radiate outward, acting as a reminder that material flows that cross cultural and national borders are mirrored in the smaller acts of consumption that form part of everyday accumulation.

Jeans and Jackets

Most women have some Western-style clothes in their cupboards, although few married women continue to wear jeans and T-shirts in daily life,[21] and these clothes have generally been passed on to younger relatives. However, there are occasions on which it is acceptable for married women to wear them, in those places where conservative Delhi attitudes to dress are considerably relaxed. Short visits to hill stations such as Mussoorie and Simla allow greater freedom to mingle with the urbane Mumbai-ites on holiday, newly married Punjabi girls in their bright synthetic saris, wealthier Indians who live in other countries but are making extended visits home, and traveling foreigners. A woman might wear her favorite sari for a dinner one evening and Levi's jeans the next, so long as her family agrees. Talking to women in their homes, it was clear that they did not often wear jeans in these social situations, and many of their descriptions of potentially doing so were wishful thinking. Clearly some husbands also enjoyed the sartorial freedom of holiday destinations and bought their wives Western clothing. By contrast, Purnima, a Punjabi economics teacher whose husband's family had come from Pakistan after Partition, lived in a rather conservative joint family: she, her husband, and their two teenage sons lived in the Progressive with her brother-in-law and his wife and her mother-in-law. She made a point of saying that she would never be allowed to wear Western trousers again. In fact, now that her sons were growing up she was under some pressure to make sure she always looked respectable, wearing suits that did not fit tightly and had higher necklines. The residual jeans, sweatshirts, and leather jackets that remain in their wardrobes enable women to imagine alternative identities, but they are rarely worn.

Women who visit relatives abroad may buy clothing in these new social settings, to augment the saris and suits they take with them. They may opt for smart casual clothes and enjoy choosing among the new designs and fabrics suddenly made available to them. Back in the wardrobe in Delhi, such clothes can be a testament to the versatility of their lives, the multiplicity of their social facets, and their ability to travel the globe to fashion and create new personas. Rohini had recently visited her sister-in-law, Rupa, a doctor in New Jersey. While she was there, her brother-in-law bought her a synthetic velvet roll-neck top and matching wide-legged trousers. It was "very expensive," and she thought it very smart.

But, equally, Western clothes can represent values and lifestyles alien to their owners.[22] Rohini's husband Suresh, an industrial consultant who traveled extensively, had bought her some pink Lycra leggings in the U.S. during the Jane Fonda craze, and although they had cost $60, she had never worn them. After her return, her sister-in-law had sent her an off-the-shoulder nylon top and some polyester-spandex ski pants, with the suggestion that they would spur her to lose weight. Looking first at the outfit and then down at her rather matronly figure, Rohini laughed. These clothes were unwearable in her Delhi life and she had neither the figure nor the nerve to wear them even in the U.S. They represented a fantasy ideal which she could not fully subscribe to, let alone strive for.

Women who do not travel abroad may also have smart Western clothes, but finding a venue to wear them in can be a problem. Asha, who lived in a nearby housing society, ran a successful local clothing shop called Asha's, which was frequented by many residents of the Progressive. She had one such outfit: a synthetic velvet, stretch-knit maroon top and A-line skirt, hanging alone on a hanger in her cupboard. It had cost her husband more than Rs 2,000 in an upmarket boutique in South Extension, a significant amount (the cost of three or four reasonable-quality suits). It was sexy and figure-hugging, but not at all in the way a sari is. She was proud of the outfit because of both its novelty and the danger to her moral standing should her family see her wearing it. She said she would wear it if her husband, Prem, took her out to a romantic dinner in a smart restaurant in south Delhi, and for this reason it represented the thrill of an imagined evening out, away from the children and her conservative father-in-law, who lived with them.

Asha also had jeans and Indian-style handprinted sleeveless tops, which she combined with sneakers, rather than the low-heeled sandals she wore with saris. On her and her husband's rare evenings off when she and their two young daughters visited me in my flat, they would bring a portable tape player and Hindi film music, together with a bottle of rum, for an evening of dancing with the children in the living room, away from her rather disapproving father-in-law. On evenings such as these, she admitted that she would sometimes leave home in traditional clothes and change in the back of the minivan they had for their business. They obviously saw my home, which was not shared with any Indian family, as a Westernized destination, a touristy "quick break."

Bulbul sometimes took her child to an amusement arcade in a new suburb and invited me along. She usually wore jeans and a shirt with sneakers, and we would stop on the way home for a treat at the very new local McDonalds and a visit to the newly opened supermarket. These outings were to new spaces within the city, where new personas could be created in order to participate in new experiences, and the social norms of dress could be reformulated for a few hours. But once back in the Progressive, she would habitually change back into a *kurta churidar* to visit the local shops.

Finally, donning Western clothing can also be a way of retreating to the freedom and security that women felt before marriage, during their college or teenage years. Visits to their family, away from in-laws, may entail mornings and afternoons cooking, chatting, and gossiping within the house. Skirts and dresses are once again as acceptable as old suits and saris, and familiar favorites are retrieved from trunks and cupboards in childhood bedrooms.

Gifting, Stockpiling, and the "Recycling Drawer"

In addition to the excesses of silks, unwanted synthetics, and unwearable Lycra leggings, the sheer volume of continuous gifting leaves women with more saris than occasions to wear them all. Trips home, visits to relatives, festivals, and personal celebrations can produce overwhelming quantities of clothing, and distant relatives are sometimes obliged to give gifts, regardless of whether they are to the receivers' taste.[23] Color schemes and styles that are "not quite right" are relegated to spare rooms and trunks. Usha opened up her cupboard of "extra" saris, which contained at least 150 cotton garments. She explained that her in-laws in Bengal gave her a sari for every Durga Puja, and she was constantly attending family weddings where she would be customarily gifted another sari. She remembered a Bengali custom in which such extra cotton saris were lent to houseguests for the duration of their stay, a welcoming gesture that literally clothes the guest in the apparel of her hostess and brings her into the realm of the family.

Responsibility for gifting can be onerous for a woman. Sangeeta was particularly organized:

> I also keep a whole set of saris in a chest of drawers for gifting away—now my father-in-law is older he asks me to arrange all the gifts for his household staff, and I generally buy up some. I get friends to buy them when they visit regions, and I pay them back. I keep all the tags on and know the prices, so I can always decide the appropriate monetary value of any gift. I have a whole range depending on what is needed.

She also admitted that she had a separate group of unwanted saris that could be surreptitiously given away at a good opportunity. These are kept physically and conceptually separate from the "extras," and it is important to keep track of where they came from, as she explains:

> Under my bed is a trunkful of saris that my mother recycles—new ones that she is gifted and does not want, or new ones she buys as and when, for when the need arises to gift someone. You have to be careful that gifts from your mother's side go to your father's side . . . [they contribute toward] stockpiling to fulfill obligations at forthcoming weddings and so on. If you are gifted a sari from someone you do not really like and do not like the sari you might give it away unworn—but once a woman found out that I had done this, and she was really insulted. I feel very bad about it.

Women were extremely reluctant to admit that they would ever give away a gift; Sangeeta was unusually open about it, but she was sure that most families engaged in similar practices. Doing so is considered thrifty if a woman can be certain that neither the original giver nor the receiver will ever know. Women sometimes hide such practices from their husbands, who may disapprove; several men claimed that if a person was thinking of you when they gave a gift, it was wrong to pass it on. Failing to incorporate a gift into the wardrobe can be a hurtful rebuke to the giver, and passing it on can reflect badly upon a family's ability to buy new gifts. While women cleverly conserve household resources, men are often concerned to maintain outward appearances, and do not want their wives' strategies discovered.

Worn-Out Things

As everyday clothes age with wear, they become less formal, transforming into second-bests for wearing locally. The least presentable garments may be the most comfortable and are reserved for wear at home. In hot summers, old, soft cotton saris are favored and women always keep a few worn ones for this purpose. The final category of clothing is that which no longer has even a marginal place in the wardrobe and which has been sorted out to be got rid of, usually everyday saris and suits, as well as men's and children's clothes. Despairing sighs over quantities of old clothes reveal how problematic discarding clothes can be.

Silk saris become stained, the material frays at the seat, and folds tear. Cotton clothing fades and wears out. Waistlines expand, fashionable garments appear dated, and old, classic silk saris are pushed to one side by newer additions. Such clothing begins to occupy a marginal zone, and takes on a liminal status within the wardrobe both physically and conceptually. Clothes could move in and out of the unwanted category; recalling their associations sometimes made women decide to hold on to pieces a little longer in case a use could be thought of for them. Clothing can lie dormant in the wardrobe, replete with latent value. It can also be recategorized several times, revalued on different criteria until a more powerful reason to alienate it is discovered. Seasonal wardrobe changeovers, such as at the festivals of Holi in spring and Diwali in autumn, provide occasions for reevaluation.[24] In particular, Holi, the annual festival of color that heralds the spring, is a time when everyone wears their oldest clothes for the traditional play-fighting with colored water before throwing them out and changing into new garments. Other opportunities include moving house, marriage, and eventually death. Often such clothing is wrapped up separately in bundles and kept in a storeroom or spare room, waiting for a suitable opportunity to get rid of it.

The Work of Self-Expression

Pressures to conform in matters of dress are reinforced through the predominance of gifting early on in a married woman's life, resulting in a lack of options for self-expression. During their lifetimes women try to achieve greater control over their wardrobes through various strategies, exercising more choice and feeling able to recategorize garments in order to bury them out of sight and out of mind.[25] They may select their own gifts to be given to them by husbands and family, or make their preferences known; they may not wear clothing they have received, or may keep clothes they are not supposed to wear. Eventually, they may be able to decide which to treasure and which to discard.

With cupboard doors pulled open, piles of clothes heaped in disarray on the bed, and suitcases spilling their contents onto floors, there would often be a respite for a cool drink or a cup of tea and a pause for reflection. Sangeeta looked around her, commenting, "Most of it displays a lack of choice, but I am really proud of the ones I have chosen myself." Having married out of her regional caste community, she was often given Bengali clothing by her in-laws, and fewer north Indian clothes than she might have expected. She had not had many opportunities to assemble a wardrobe herself, and had not had a lot of money for many years. Most of her clothes had associations overdetermined by family histories, events, places, and people. Sangeeta was rather sad that her husband had not personally given her very many clothes. Perhaps, she admitted, she had criticized his taste early on in their marriage. The wardrobe can represent members of the family and close friends, who have provided its contents through gifting, inheritance, lending, and swapping. It can also reveal gaps where clothing should be, rather like a family gathering when an important member is absent. There were clothes that should have been given by significant others, ought to have been worn, kept, and handed on.

In contrast, Rohini voiced for many the dilemmas of opportunity. "[There is] such a large choice of clothing now, it is overwhelming; saris, suits, and now the whole range of Western clothes coming into India." The scale of women's collections was often rather daunting, perhaps even embarrassing, when they were presented in their entirety to an outsider. As we went through her clothes, Rohini often admitted that few of them were her own choice. But she also said that looking in the wardrobe made her feel good that so many people had thought of her: "I do not *have* to go shopping for myself." Over the years she had developed a particular liking for pinks in all shades, and her close family had come to know this; her favorite was a south Indian block-print sari with a seven-dot pattern in pink, given to her by her brother. As a young woman she had wanted pastel chiffons (in contrast to her trousseau of rich silks, which she disliked at the time), but over the years what she was given, what she wished to wear, and the sentiments that she felt for the givers of the clothing had gradually harmonized. Rohini had trained in economics and computing, but now kept house; she had two teenage children, a boy and a girl. Her sense of herself as a woman with certain tastes grew as a result of her both reaching her mid-thirties and being increasingly brought into the fold, wearing what middle-class wives and mothers wear, through receiving saris of a certain type. She was able to influence those gifts with her developing confidence and individuality as she matured.

Wardrobes as Collections

The Godrej cupboard plays an important role as mediator between its contents, the internal set of possibilities, and the external world of performance. Trunks, tough cases, and above all the Godrej cupboard are the ideal type of storage: they are solid, permanent containers. The hard grey surface of the metal container exists in contrast with the dazzlingly bright soft silks and cottons revealed within. There is a homology between the body, clothing, and wardrobes, all of which articulate the tensions between interiority and exteriority from the center of the person outward. Clothing covers the body, and the wardrobe contains the clothes; the wardrobe is like another layer of skin on the person, and is itself contained within the home, the realm of the family and the domestic sphere. Through the manifestation of the relational self, made visible in the clothing within, the wardrobe relativizes concepts of inner and outer persons; tensions between individuation and relationality operate at every layer.

The cupboard full of clothes is usually located in the marital bedroom, and it itself is the product of the marriage; it is person-like through its relationships with members of the family, and can be spoken of in these terms. The wardrobe is enhanced by the relationality of the clothing contained within it; it provides a "body" containing the "second skins," the exuviae that animate it and give it its efficacy. In this sense, it is analogous to the Tahitian A'a figure discussed by Gell, which images the notion of personhood as the aggregate of external relations (the outcome of genealogy) and at the same time the notion of personhood as possession of an interior person (Gell 1998, 137–39).

The constant flow of clothing into the home, where it is incorporated into the wardrobe and expelled once again, fuels growth and change; the wardrobe is a conduit through which the dynamic relationships between subjectivities and objectivities, the facets of the self and social relations, are constituted and made visible. It is a stopping-off point for a piece of cloth as it travels through networks over time; it may be conceived of as a container, a collection, a place in which to sort, hide, rediscover, or reject: it marks the pinnacle of a hierarchy of domestic spaces where clothing is constantly evaluated.

The management of clothing in the wardrobes discussed here reveals many of the characteristics identified with the practice of collecting (Baudrillard 1994; Belk 1995; Pearce 1992, 1994, 1995). Though they start out as fairly passive recipients, brides-to-be, many women are lured by the "thrill of the hunt" for an elusive sari and begin to search for unique pieces. The act of incorporation into the wardrobe is akin to a possession ritual, reenacted through wearing, laundering, folding, and storing, turning a commodity into a singularized object. "Control, one way or another, is what makes an object become more a part of the self" (Pearce 1992, 55). Collections are often interpreted as extensions of the self, which can be manipulated and rearranged because of their inherent seriality (Baudrillard 1994). Indian women's wardrobes are dynamic spatio-temporal extensions of the self that engage directly with their sense of connectedness to others. It is the active organization of clothing that turns its accumulation through social roles in early married life into the building of a collection in later life through processes of detachment and reorientation.

Just as the "material value of the souvenir is an ephemeral one when juxtaposed with a surplus of value in relation to individual life history, so the ephemeral quality of the collected object can be displaced by the value of relations and sheer quantity" (Stewart 1993, 166). Thus begins the process of attachment and detachment that enables unwanted pieces to be rejected from the collection. Things stop being attached to the person and start to become attached to each other; getting rid of one item implicates the whole network of people and things in which it is embedded. The wardrobe, is in this sense, a distributed mind, whose composition can be altered and manipulated through the inclusion and exclusion of its parts.

Recent work on ridding in the UK has shown that disposal is not a sign of a throwaway culture, but rather should be understood as practices of saving and wasting that materialize relational social identities within the domestic sphere.

> Acts of getting rid of things . . . are actually acts that are critical to the performance and regulation of the self and a fundamental part of identity work. Indeed . . . to throw away (certain sorts of) things is an intrinsic part of contemporary being; a way of narrating ourselves through the presence and absence of consumer goods. (Gregson, Metcalf, and Crewe 2007, 688)

Similarly, with Indian practices it is the degree of relationality in which these clothes are bound up that matters, and the impact this has on decisions to cast out things. The feeling of being loved and the reaffirmation of ties which envelop a woman within a network of relations is finely balanced against a desire to express individuality, growth, and change, a desire that each woman fulfills more or less successfully through her lifetime. Some clothes are treasured for its sentimental value and remain as souvenirs within the home, but we will also see how some clothes no longer "work" within the synchronous collection; women detach themselves from certain items, yet may reuse their relationality to increase their social value.

FIGURE 4. The local *presswala* ironing domestic clothing at a pavement stall outside the Progressive Housing Society. PHOTO BY LUCY NORRIS

PLATE 1. The Ghora Mandi wholesale second-hand clothing market at Raghubir Nagar.
PHOTO COURTESY OF TIM MITCHELL.

(above) PLATE 2. A dealer in recycled *zari* threads in the Kinari Bazaar, Old Delhi, displaying sari borders and a heavily decorated skirt.
PHOTO COURTESY OF TIM MITCHELL.

(right) PLATE 3. Burning a silk sari to extract silver thread, Kinari Bazaar, Old Delhi.
PHOTO COURTESY OF TIM MITCHELL.

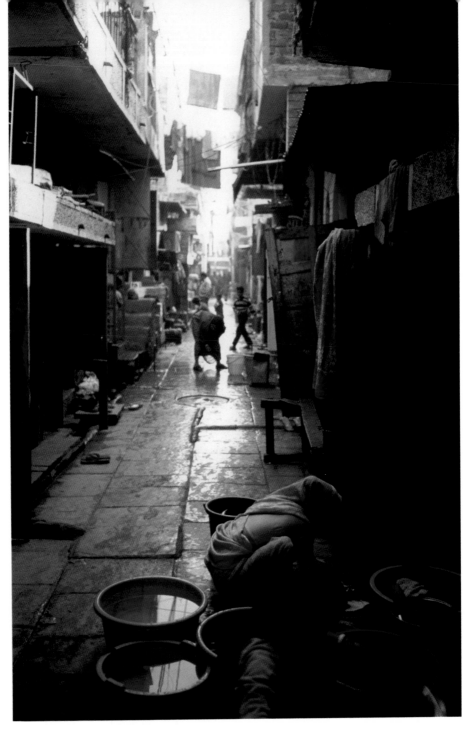

PLATE 4. Washing old clothes in an alley in Raghubir Nagar; they will later be resold in the Sunday second-hand markets. PHOTO COURTESY OF TIM MITCHELL.

(below) PLATE 5. A typical Waghri household displays their own *bartan,* glass, and ornaments on shelves. PHOTO COURTESY OF TIM MITCHELL.

(above) PLATE 6. A Waghri woman sits behind her clothes for sale in the Ghora Mandi. On the wall behind her a wholesale dealer has piled up bundles of trousers he has bought that morning. PHOTO COURTESY OF TIM MITCHELL.

PLATE 7. Old *dhotis* and saris are torn into squares for sale as wipers at a Muslim wholesale shop in Azad Market, Delhi. PHOTO BY LUCY NORRIS.

(above) PLATE 10. Detail of a patchwork bedspread showing animals, paisley motifs, and other design elements from twenty different saris.
PHOTO BY LUCY NORRIS.

(top left) PLATE 8. A Waghri woman washes clothes for resale in a side street.
PHOTO COURTESY OF TIM MITCHELL.

(bottom left) PLATE 9. A man takes a break from ironing clothing; two tailors are repairing clothing in the background. PHOTO COURTESY OF TIM MITCHELL.

PLATE 11. Waghri women selling second-hand clothes on the pavement in Panipat, Haryana. PHOTO COURTESY OF TIM MITCHELL.

(above) PLATE 12. Stocks of old silk saris and *lehenga* (skirts) at a dealer's house in Raghubir Nagar. PHOTO BY LUCY NORRIS.

(right) PLATE 13. Bedspreads and cushion covers made from silk saris are sold alongside recycled embroidery at the Gujarati market off Janpath, New Delhi.
PHOTO COURTESY OF TIM MITCHELL.

(above) PLATE 15. A recycled sari coat for one of Mrs. Sharma's overseas clients. PHOTO COURTESY OF TIM MITCHELL.

(left) PLATE 14. Mrs. Sharma inspecting a delivery of second-hand saris to be included in an export order. PHOTO COURTESY OF TIM MITCHELL.

PLATE 16. *Sanskar by Sonam Dubal 2001* on the catwalk, Delhi.
PHOTO COURTESY OF POMY ISSAR.

4 LOVE AND PROTECTION
Strategies of Conservation

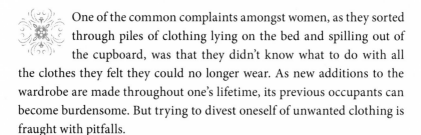 One of the common complaints amongst women, as they sorted through piles of clothing lying on the bed and spilling out of the cupboard, was that they didn't know what to do with all the clothes they felt they could no longer wear. As new additions to the wardrobe are made throughout one's lifetime, its previous occupants can become burdensome. But trying to divest oneself of unwanted clothing is fraught with pitfalls.

Cloth is supremely capable of articulating a multiplicity of ideas and values, and women trying to divest themselves of clothing assess these ideas and values through a series of rationalizations. Cloth can make connections, wrap and bind, and transmit the essences of people and places, and this chapter discusses how these capacities are conserved within the domestic realm wherever possible. Daily practices reveal popular discourses concerning the moral virtue of thrift, the strengthening of familial relations, the importance of sociability and wholeness, and associated concepts of hierarchy, purity, and pollution. Part of the technology of everyday life, cloth is a social operator, and the materiality of cloth is a resource to be utilized; clothing can both protect and endanger, and transactions involving it affect both giver and receiver. Cloth is ephemeral, but recycling can prolong its life whilst renewing its form. Cloth mediates spatial and temporal relationships as families and individuals re-create themselves anew.

The sari and the *dhoti* are just lengths of cloth. They lie flat but can be folded, wrapped, draped, or cut up; they are fundamentally adaptable to any wearer. In the domestic Indian context cloth's material physicality impels it to be used and reused, over and over again, and this necessity is reflected in women's comments, attitudes, and actions; it is actually very difficult to decide that a sari is worthless or no longer useful. It may have no immediate or apparent use, but it usually has potential. Once bound up with the ecology of the wardrobe, cloth also forms part of a social environment; it is a material agent in the formation of subjectivities and social relations. Drawing on the work of James Gibson (1979), Ingold writes, "Meanings are not attached by the mind to objects in the world, rather these objects take on their significance—or in Gibson's terms, they afford what they do—by virtue of their incorporation into a characteristic pattern of day-to-day activities . . . Meaning is immanent in the relational contexts of people's practical engagement with their lived-in environments" (Ingold 2000, 168).

Priya's Waistcoat

Priya is a widow in her eighties, originally from Kumaon in Uttar Pradesh. She used to live in a well-to-do suburb in south Delhi until the expense proved too great and she moved to the Progressive to live with her daughter, who is divorced. Her former son-in-law lived in Canada. A trained artist, Priya is known for her highly unusual and beautifully coordinated clothes, which are characteristically asymmetrical in pattern and do not "match" in the conventional Indian sense.[1] They incorporate her native Kumaoni tailoring into classic Delhi styling.[2] She was married in the late 1930s, at eighteen; her father was a senior officer in the police force, and they were living near Banaras. Looking though old photographs of her family, she remembered that by the age of fifteen she had already started to wear her own style. In the pictures she wore the small fitted waistcoats that were to become her trademark. These were worn on top of her *choli,* with the end of her sari draped over the top. Her tiny figure has remained the same all her life, so she has never grown out of her clothes. When I spoke with her she generally combined the waistcoats with a *kurta* and *churidar.* Her two favorite waistcoats were both extremely fine-quality items from her trousseau: a black velvet waistcoat from her sister that she still wore every winter, and a salmon-pink silk one from her parents. The silk was figured with flowers and leaves woven with silver thread as a supplementary weft (*zari* work), in the "heavy" Banarasi style. A highly prized item of clothing in Priya's wardrobe for over sixty years, it was in an unusual style that could only have been worn by an upper-class woman for special functions, and which Priya had appropriated for her everyday wear. The waistcoats thus represented her youth, her high social status, and her reputation for stylish dressing.

However, this old waistcoat was now long past its best. It was stained with sweat on the inside and had torn along the weft in several places, probably as a result of folding during storage. It was frayed around the edges and looked generally tatty, past repairing. As a thrifty old lady trying to divest herself of unnecessary items at the end of her life, she had occasionally wondered what to do with this garment. It was not sufficiently unique to be restored and sold as an antique piece. She had no inclination to recycle it into smaller domestic items, nor had her only daughter. Priya had tried to give it to her maidservant, but as it was torn and dirty she had refused it. Priya could not think of anyone else to give it to. She seems to have thought the *zari* waistcoat was worthless, until an acquaintance in the Progressive suggested that she might be able to get something for it if she sold it by weight in the old street of goldsmiths off Chandni Chowk, in Old Delhi. For the smiths, its value lay in the silver that could be reclaimed from it rather than in its usefulness as clothing or in the labor invested in the *zari* work. Hoping to "make something from it," Priya gave it to her acquaintance, who said she went there often to deal in jewelry and had hinted that the waistcoat might fetch as much as Rs 300. Six months later, however, nothing had come of it, so Priya asked for the waistcoat back.

She then thought about approaching the *bartanwale,* but she was reluctant to ask them into her house as she had not done this before and was nervous. She said she did not know what to expect in return for the waistcoat, but wanted to get a non-stick *karahi* (frying pan) for it; this semi-luxury item was what she really wanted in her kitchen. Such a utensil would be available in the local shop for perhaps Rs 200–300, depending on the brand. Bulbul, who had introduced me to Priya, offered to do the bargaining for her, together with Uma, who had dealt with the *bartanwale* before. Uma therefore waited until she saw one of the *bartanwale* sitting in the road outside the Progressive and made an appointment to meet her the next day at Bulbul's flat.

The *bartanwala* was a young girl called Mira, who looked distinctly disappointed by the waistcoat and offered Rs 50 for it, in cash or as part payment toward a *bartan.* With the figure of Rs 300 in their heads, Uma and Bulbul were somewhat surprised, but Mira laughed and repeated her offer. A half-hearted, nervous bargaining began. The fineness of the *zari* work was pointed out, but Mira was unimpressed, and said that there was no market for this. The old clothing was sold on to the poor, and poorer people could not afford to pay extra for handmade *zari* work; they needed practical, wearable clothing. She claimed no knowledge of the value inherent in the silver thread or the trade in metal thread. Indeed, she countered that if the piece had been decorated with synthetic *zari* it would have been worth more to her. It would have been shiny, lasted longer, and been easier to keep. Also the fact that it was a waistcoat reduced its value— perhaps a bit of real *zari* on a suit or sari would have added value, but no-one could afford the luxury of a tatty waistcoat, which is not an essential garment. Neither Bulbul nor Uma really knew what the second-hand value of the garment was, which undermined their confidence.

Since Priya had her heart set on a new, modern frying pan, a long discussion ensued about what size and quality of pan could be had. But Mira would not trade a *bartan* for this waistcoat alone, and asked if they had anything else they wanted to get rid of as well, such as other clothes, electrical goods, or toys. The bargaining was wound up inconclusively, with a sense of disappointment all round. Later on, however, when Priya discovered that she could not get the frying pan unless she had something else to trade, she was extremely upset and, in tears, cried, "Just throw it away." The plaintive cry was an expression of frustration and disappointment that her waistcoat was worth no more than rubbish. At that point I suggested that Bulbul and I try to sell it for the silver ourselves.

Eventually, we made the journey into the narrow lanes of Shahja-hanabad to the Kinari Bazaar, a wholesale market for wedding and other festival decorations; gold- and silversmiths have traded there for centuries. Amongst the dealers selling antique jewelry, wedding ornaments, and bangles were craftsmen selling rings, necklaces, and bracelets fresh from the small hearths tucked away in the backs of their workshops. Nearby were two small hole-in-the-wall shops, one with a tatty old *ghagra* (a full skirt with *zari* work) hung up outside as an advertisement. The shops contained large old balance scales, a hearth, tongs and crucibles, and all the paraphernalia for burning clothing (plates 2 and 3). The process leaves strips of blackened silver wire to be transformed into tiny new beads of shining metal. These are then sold on to the few families nearby who continue to manufacture *zari* thread by hand.

We approached both dealers with the waistcoat. First the threads were rubbed against a touchstone, and nitric acid was applied to the streak to determine the metal and its purity. Next the waistcoat was weighed in the scales, and the percentage of silver thread calculated. We were offered Rs 250 by one dealer and Rs 300 by the other, showing that selling it in this way was the most profitable means of recycling this garment available. Eventually, I myself paid Priya Rs 300 for her waistcoat, and brought it back to England as part of a collection of recycled clothing I had been commissioned to make for an ethnographic museum. However, it was not included in the final selection, because it was too atypical in style. It was eventually put on show in the foyer of the Department of Anthropology at University College London as part of a display about Indian recycling and value, its own journey highlighting the curious resonance between discards and museum collections, both being items that have been re- moved from circulation (Pomian 1990).

Hand-Me-Downs

Family Envelopes

Often the trajectories of clothing bring the broader cycles of time and ongoing relationships to mind, rather than particular events such as rites of passage. As Rohini put it, "When I see these clothes it makes me remember certain people: they are not transitional gifts for the moment, they last and stay in your wardrobe." Saris, suits, shawls, jackets, and unusual pieces can all be cut up and remade as they circulate through the generations. As women move through the life cycle, their clothing moves in the opposite direction: it passes down through the family as women get older and divest themselves of pieces in order to enfold the next generation. The process begins with the birth of a new baby in the family, continues through childhood, marriage, and motherhood, and is completed at the time of death.

A loved one's used, worn cloth is often passed on to another person, offering comfort and protection and instilling the original owner's spirit in the recipient. When the previous wearer is considered to be a particularly good person, admirable and morally upright, the most positive use for such old cloth is for protection of the vulnerable. At birth, babies are wrapped in quilts made from torn-up pieces of old *dhotis*. Since the most dangerous period of the life cycle is thought to be that following birth, extra care must be taken of new-born babies. For the first week or so, they should be wrapped in old washed cloth, such as soft old *dhotis,* which should be white; these cloths are also used to make padded diapers and baby quilts (*dupta*).[3] Several women said that the prospective grandmother would make them, using the family's old clothes or those of "auspicious ancestors." Sangeeta related, with some pride, a story of how a relative's baby was in need of particular protection after a very difficult birth. The baby's mother asked for one of Sangeeta's father's *dhotis* to wrap it in, her father having been considered a particularly fine, upstanding old man. Likewise, young children who are ill may be wrapped in someone else's clothes, such as a family member or a neighbor who has enjoyed a long, auspicious life.

Related to this practice is the belief that new clothes make the child look attractive and could attract the evil eye (*buri nazar*). It is not considered good to attract compliments: some people might be envious and speak with a "black tongue" (*kali zaban*).[4] Children are further protected through charms, necklaces, cords, and kohl. Although many urban mothers, particularly working women, are increasingly using disposable diapers, and making diapers by hand is often thought to be too time-consuming, most women stressed that the first few days after birth still required the extra protection handmade ones provided. Mothers and mothers-in-law enforce these beliefs. Madhuri, a university college teacher in her late thirties, recounted how her mother had taken her first grandson, Prakash, to the temple as a baby, wrapped in the dullest, oldest cloth to avoid attracting the evil eye.

Janaki related the story of her brother's baby dress, which was made by her mother for him on his birth in the early 1960s. She herself was dressed in it the following year; then her son Arun wore it. On my arrival in May 1999, one of Janaki's neighbors, Divya, had just had a baby. Divya was an academic researcher and teacher, a Bengali Brahmin who married an academic from Uttar Pradesh. Her father was a doctor who had traveled extensively, and her parents lived near the Progressive with her younger brother. Another sister lived in the U.S., and the sisters visited each other regularly. Divya and Janaki were close friends, and Divya's parents were also old friends of Janaki's family. So the dress was duly handed on. Although by now it had become a bit torn and holey, "it was fixed up with ribbons and lace to look nice." Divya is expected to hand it on again if it is still in good condition. The auspiciousness of Janaki's family was bound into the fabric of the dress and wrapped around the baby.

As old cotton saris and *dhotis* are repeatedly wrapped around the body, washed, and worn again, the dyes fade, the fibers soften, and the threads wear thin. Many children's clothes remain in fashion as everyday wear for a long time, and in large extended families aunts may be close in age to their nieces, so clothing passed on from a young aunt may not be very old. But whether clothes are inherited from parents, aunts, cousins, or siblings, some achieve a special status within the family over time, especially if they had an initial high value as exotic Western high-quality goods brought into India by traveling elders as souvenirs and gifts. Madhuri's family seems particularly adept at keeping clothes in the family. Her mother was a non-Brahmin Bengali and her father was a Maharashtrian Brahmin from Bangalore. He had brought Madhuri back a red anorak from Japan when she was young; now her daughter and son have been wearing it. Madhuri's husband, who was from Tamil Nadu, was a software engineer who was often posted abroad; they had lived in the U.S. until returning to the Progressive. Some of their clothes travel extensively:

> This is a small sweater that originally came from England; it's really good quality, and has a long history. I have a close cousin in Kolkata, and it was her husband's; he had given it to his cousin's brother, who then gave it to me. I then gave it to my brother, who took it to Australia; he then gave it back to the Kolkata cousin's daughter. It then came back to my son Prakash, but as I could still fit into it, I started wearing it again. Eventually it has worn out; now I have passed it on to Hema [her live-in maid].

Clothing for younger girls includes a variety of Indian styles for special occasions, and often these are made from old saris that belonged to their mothers and grandmothers, many with sentimental significance. For her younger brother's wedding, Asha had her tailors cut up a special *bandhini* silk sari from her own trousseau for her daughter Ria, and another plain green satin one for her younger child, Padma. The decorative end of the sari, the *pallu,* is usually used to create a bodice panel for the *choli,* and both bodices and skirt hems can be newly embroidered so that both pieces match and look as though they were made as a set. Similarly, Uma's mother gave a *chaniya choli* set (a long blouse and skirt) made from a favorite sari to her granddaughter Shashi on her first birthday. It was so long that it had three huge tucks in the skirt, with about twenty-four inches of material in each one; Uma keeps letting them down, so that Shashi can still wear it although she is now nine years old.

Many older women believe that children do not warrant expensive new clothing, as they are not yet social adults; such spending is more appropriate once a child has grown up and married. The examples discussed here show that reused cloth has great sentimental value; it folds young girls into the family by draping them in the fabric of their female relatives, imbuing them with the good qualities of their elders as they grow up.

Much younger girls also often wear very fancy, European-style synthetic party frocks, made of brightly colored shiny satins, with frills, lace, and deep ruffles, embroidered with flowers and imitation pearls. Usually worn to family occasions and special functions such as weddings, these dresses themselves can be objects of desire and are passed between friends, colleagues, and relatives. They are often very expensive if locally made, even more so if imported, and are thus markers of status and distinction. They travel through networks of middle-class people, often following paths that are mapped out two or three children ahead. For upper-middle-class families steeped in a tradition of sartorial modesty, dressing up children, especially girls, seems an acceptable display for parties and one which everyone enjoys, even though those clothes are usually more shiny and flashy than adults would consider wearing.

Sangeeta showed me a very frilly white dress, which had been given to her by the mother of a friend. Carefully wrapped in tissue and encased in plastic, it had already passed through several hands, and now was waiting for her daughter to grow into it. Once she finished with it, two more cousins of Sangeeta's had younger daughters eagerly waiting in turn, and it was acquiring a good deal of sentimental significance along the way. Madhuri had kept her own "foreign" dress from her first birthday; her daughter Pia had then worn it, and it had now passed to a cousin's daughter. Madhuri had just been gifted with another family party dress, which Pia was waiting to grow into; she was expected to wear it for her eighth birthday. Even Manju, an unmarried woman in her twenties, had kept her own favorite frilly dress in her cupboard, hoping to have someone special to pass it to one day.

These ready-made clothes are still highly prized by contemporary mothers, whose own mothers had valued the luxury of shop-made or imported fancy dresses at a time when most clothes were made at home or by the local tailor. But as Madhuri pointed out, since clothes are increasingly bought in the market ready-made, it is now the handmade gift from a person close to you which is treasured, especially clothing made from significant pieces of material.

Learning to wear clothing through dressing up and playing out roles is an apprenticeship, a way that a girl prepares for adulthood. The performance of wearing such clothes at gatherings also seems in turn to be designed to bring the older generations back into the familial fold through emotional reminiscence, confirming relationships and bonds of affection, as Madhuri made plain regarding her eight-year-old daughter Pia:

> Pia has a child's sari, which I was given when I was eight by my aunt. My aunt married in 1952. She cut the border off one of her wedding saris, sewed it onto a piece of plain cloth, and made it look like a sari for me when I was a little girl. Pia recently wore it for her grandfather's seventy-fifth birthday, as his sister, the aunt who had made it originally, was also at the party.

Likewise, girls often wear Punjabi suits that have been handed down by older relations or made up from old saris. The saris are cleverly cut up, with the *pallu* used to make a central panel down the front of the shirt, or upper and lower sari border pieces are sewn onto necks, hems, and cuffs, or onto a plain *dupatta*. Prints and weaves with patterns that get denser toward the *pallu* are arranged so that the denser part appears toward the bottom of the *kurta* or the *salwar*, and stains and holes can be carefully avoided by a master tailor when cutting out panels. Reusing cloth in this manner is somewhat risky—should the tailor make a mistake, the original sari is lost and a memento that has been kept for years is destroyed. Such bad workmanship is particularly lamented, and good tailors highly praised.

Madhuri wanted to cut up her wedding sari, a nine-yard Maharashtrian that would make a "new" six-yard sari, leaving three yards for a dress or tunic. But although she has done this with others, she has been told that it is unlucky to cut up the sari one was actually married in. For some men and women, cutting up saris is a very touchy subject. The morals of thrift must be weighed against the belief in the auspiciousness of the uncut cloth worn for a holy ritual. Other people are simply unaffected by either concern, while a few positively reject such religious associations. Sangeeta's sister, now resident in the U.S., brought her young son back to India for his sacred thread-tying ceremony (a Hindu initiation ritual for upper-caste boys), for which he wore a new *dhoti*. Sangeeta had had the cloth made up into a *kurta* for her sister. Not at all moved by the ritual significance of the clothing and attaching no sentiment to it, she had left it behind on her return to the U.S.

Sharing and Swapping

Female relatives may swap saris around as they get bored of them, and the clothes become expressions of shared taste and affectionate relationships.[5] Saris can be worn by women of any age, and grandmothers, mothers, and daughters can pick and chose as the fancy takes them, whether asking outright for one or just hinting. Once a woman marries, her circle of female kin is instantly extended. Not only does she receive old clothes from her mother, aunts, and cousins, but as she gets to know her new family, she may start to be included in their constant swapping as well. Sangeeta's relationship with her mother-in-law had been good: one of her favorite saris had been passed on when her mother-in-law saw her "drooling over it."

Used clothing given in affection can enfold its recipients more intimately than an obligatory gift, but sometimes things are passed on quickly as nobody likes them. One sari had acquired significance simply because of its long history, passing in and out of families connected through blood and marriage like a hot potato. Madhuri unfolded it to show me: a pink silk sari with a border decorated with an elephant motif. Her mother-in-law had given it to her after receiving it from her other daughter-in-law, who had been given it by her mother, who in turn had been given it by *her* mother. Madhuri said that it had been passed on so much because none of the husbands had liked the shade. Neither did her husband, Alok, but she wears it sometimes anyway, as she likes it, and she will not just "give it away" (meaning to a non-relative), but will keep it for her daughter Pia.

Women remember their grandmothers and great-grandmothers with fondness when shaking out a folded sari that was handed on, naming its type, describing its designs, and recounting who had gifted it to whom and on what occasion. Often these saris are not smart enough to wear without salvaging, and women are creative in their attempts to restore something to a wearable condition through imaginative patching or remaking. Rohini had a silk sari handed down from her husband Suresh's grandmother, a thick, vibrant blue color, completely plain but very fine quality. To dress it up and conceal a ragged hem, Rohini had an *ikat* border from Orissa sewn onto the bottom.

Many women wear favorite, tatty old saris around the house to "feel good." Wearing a "lovely old sari" can make a woman feel as though she is literally wrapping herself up in the folds of a maternal relative. Jaya, a widow in her late fifties who worked as an administrator for an international news agency in New Delhi, had recently lost her mother, and now had many of her mother's old silk and cotton saris in her cupboard. She often came home tired from work and changed into them in the evenings, enjoying the pleasure of relaxing in clothes she remembered her mother wearing in the past, without worrying about her appearance.

Most women seemed to be happy to accept clothing for themselves from friends and relatives, but two women independently mentioned that if their husbands asked, they might pretend the clothes were new. Some men, such as Purnima's husband, take pride in displaying their family's status through new clothing and branded goods. A man may also wordlessly disapprove of too strong a connection between his wife and her female relatives. Purnima admitted that she rarely saw her own family, who lived in south Delhi, because of the demands made upon her time by her job and her roles as wife, mother, and daughter-in-law in a busy joint household. Her mother-in-law rarely allowed Purnima time off to visit her family, and swapping clothing with her mother or sisters was now impossible.

Aging and Preparing for Death

As women grow older, their sartorial habits change; they begin to give away clothing. Once children have grown up and married, they and their own children become the focus of family wealth and gifting cycles. Ideas of dress and behavior appropriate for older women varied considerably between families from different backgrounds, as did beliefs concerning death, such as what a woman should wear for her cremation and what should happen to her belongings afterward.

A woman's wardrobe waxes and wanes as she progresses through life. The bright colors, new silks, and richly decorated fabrics of the young married woman are gradually replaced by the paler pastels of older women. As a widow, Sangeeta's mother could not really wear bright saris, so she had passed on the reds and oranges to her daughter, and had been gifted new saris in whites, creams, and pastel shades. But Sangeeta did not wear these bold colors daily, as they were too bright or heavy with gold; her taste was different, and she wondered, "What to do with them? I cannot give them away." In addition, her mother had kept all her special wedding saris and had duly given them to her daughters. But they were frayed and torn, so Sangeeta could not even wear them to marriages.

Pieces of sentimental significance are also passed on, with accompanying stories. Sangeeta had her grandmother's wedding *odhni* (veil), which she used as a bedcover in the autumn and winter months. Dating from the early twentieth century, it was a double loom width, signifying her family's high status and wealth. It was bright yellow and covered with red handprints. In the Brahmin Kumaoni tradition, a bride's female relatives make such handprints on her veil before she leaves for the full wedding ceremony. Sangeeta herself did not have this part of the ceremony, having married "out of caste" into an intellectual Bengali family who eschewed flamboyant display. Yet she also had her own "handprinted" shawl in her wardrobe, of which she was very proud. Although her aunts had refused to make one for her in the traditional ceremony, as she was not marrying within her caste, her mother had stayed up alone half the night printing one for her, just so that she would have one for her trousseau.

Older women give away treasured pieces and surplus silks when they entertain younger relatives; the choice of gifts can depend on their recipients' marital status. Sangeeta had a special deep pink and gold sari, which she had bought, on behalf of her father, as a birthday gift for her mother. Her mother had originally wanted to be cremated in it: she believed that if a woman died as a *suhagin,* that is, before her husband, she should be cremated in a sari given to her by him. However, Sangeeta's father had died first, so her mother had packed the sari away in a trunk and had never worn it. It was inappropriate for a widow, who should wear a pale cream or white one to the pyre, so she eventually gave it to Sangeeta to wear.

Usually a woman's daughters and daughters-in-law distribute her clothes amongst her female relatives when she dies. Madhuri was her mother's only daughter. She and her only brother's wife sorted through her mother's wardrobe and gave one silk sari and one cotton sari to every daughter and daughter-in-law in the extended family and to close friends of her mother. Some of her younger nieces merely got nylon chiffon ones. Madhuri and her sister-in-law kept eight or ten silk ones for themselves. Her mother died a *suhagin,* and the clothes of a *suhagin* are considered auspicious by both Madhuri's Bengali family and her husband's Tamil family.[6] Her daughter Pia saw the distribution of all her grandmother's clothes as a three-year-old and had already chosen her favorite out of Madhuri's saris for her own inheritance. Men's clothing can be equally precious. When Sangeeta's father died, he was eighty-six. He had had a successful life and all his children were well placed, so his clothes were considered auspicious and were gifted away to people who appreciated them.

Rohini had been very fond of her husband's grandmother and had one of her Kanchivaram silk saris. She felt this was a special sari, imbued with the deceased woman's good spirit, and would wear it on an occasion that had great meaning, like a child's first birthday. Rohini expected her own mother's special saris and shawls to be willed away to her three daughters and one daughter-in-law, describing them as *nishani,* that is, something that is a part of that person that comes to you—"you feel it and you think of that person."

Usha, a Bengali Brahmin, claimed that one never gives away one's wedding sari, as the senior daughter-in-law will inherit it (although she will probably never wear it). Radha, a Gujarati, also believed that wedding saris should never be given away; but if a woman is widowed she should give all her brightly colored saris away. In contrast, a *suhagin* would be cremated in her wedding sari and the rest would pass to her daughter-in-law. Some women expect to receive clothing as a sentimental remembrance of a loved mother (either during her old age or after her death), while others expect the daughters-in-law to inherit it, but the conversations I had with women in the Progressive made clear that women could be highly strategic and circumvent expectations by choosing to hand on valued garments while they were still alive.

Sympathetic Magic

Just as clothing from an auspicious person carries this quality on to the next wearer, so pieces of cloth given as religious offerings link their givers with the divine. Those who offer cloth do so in hopes of divine protection, blessings, and good fortune. Rags and tatters are left as offerings at shrines or tied to trees, and even thread may be wound around branches and idols.[7] Women leave pieces of red and gold cloth tied to branches; some have glass bangles attached as a prayer for a good marriage, while small rag cradles contain pebbles as a fertility offering. Today rags are also hung around the rear-view mirrors of taxis and rickshaws.

The counterpart to the belief that auspicious cloth can transfer benefit and protection to its recipient is the fear that harm may befall a donor of used garments if they are misused. Susan, a Syrian Christian of the Marthoma order who was from Kerala but had been brought up in Delhi, explained that she was worried about what people might do with her old clothes, citing the example of a *pandit* (priest) curing a sick child by asking the parents to pray in front of a healthy child's cloth. The fear is that the sickness might be transferred back onto the donor, through the principles of sympathetic magic. For this reason, she gives her old clothes only to her servants, whom she trusts, and only one at a time, so she knows they will use them up and not pass them on to strangers. Another woman confirmed this concern. Since boys are of primary importance in the family, she said, their cast-offs are not just given away to anyone, for fear that they will be used in this way; the threat was not considered to be so great to women and children.

Many women feared that their old clothes were being used by servants or strangers as menstrual rags, and in fact poor women do buy old saris sold by the Waghri dealers for this purpose, costing about Rs 5 per sari. An upper-class NGO director working with poor women in Delhi estimated that at least 80 percent of middle-class women who traded their old clothes to the Waghri did not realize this, and would never do it if they knew.

The belief that harm can befall a person through the perceived misuse of their old clothes can be linked to a more general set of traditional beliefs concerning *tona,* a type of black magic. *Tona* allows its practitioners to cast spells on an unfortunate victim using products of the victim's body, such as hair or nail clippings, or cloth that, like skin, is shed from the body. Clothing is therefore both a part of the person and an image or likeness of them. As Gell notes in discussing volt sorcery, this elides the Frazerian distinction between contagious and sympathetic magic (Gell 1998, 103–104). To protect against such malignant forces, some people thought that one must gather up children's hair and nails and wash them away completely, never leaving hair on combs or brushes where it could be found. Similarly, clothing worn next to the skin should be handed on carefully: it can be used to protect a trustworthy person who might have need of it, but must never be given to someone who may have ill intentions toward you. Such beliefs were rarely discussed openly with me by women in the Progressive, and although references were often made to "misuse" of old cloth, only Susan directly confirmed that she was concerned about black magic, couched in vague terms.

Reuse and Recycling

Patchwork and Quilting

Expensive pieces of highly prized fabric are often treasured and patched together as clothing and furnishings.[8] Thrift is an important factor in patchwork, but the medium of cloth allows for the manipulation of beliefs concerning protection, connectivity, and wholeness.[9] Patchwork protects its wearer if the cloth originates from auspicious members of the family. The robes of *jogis,* a mendicant caste of itinerant healers who are occasionally also snake charmers, also use patchwork. Traditionally *jogis* wore quilted robes, *chogas,* which were made by their mothers from tatters of old clothes sewn together and layered. These quilted waste pieces, called *gudari,* were then dyed saffron, the mendicant's color.[10] The baskets and wrappers in which the snakes were kept were also made this way, and topped with a cowrie shell, a charm against snakebite. The use of rags and tatters is common amongst religious beggars, ascetics, and renouncers of the world, who rely upon donations of clothing and food. Both Sufi and Buddhist monks are renowned for their pieced patchwork robes. "The patched robe of the Buddha or of a saint belongs to him in his nature of Saviour. The rags are given a new wholeness. They clothe holiness" (Kramrisch 1989, 79).

Quilting is still done at home in contemporary Delhi, though it is perhaps less commonplace amongst the generation of mothers today. Several women had had their old saris made up into quilted items in small local workshops in the market. These tailors would normally buy polyester wadding on a roll and reuse the silks that were brought in as required, making lightweight bedcovers for winter, shoulder bags, and cushions.

Madhuri had continued her Bengali mother's family tradition of making babies' quilts by removing the borders from old saris, tearing the remaining "field" up into pieces about two meters long, and folding these into pads. She would then stitch the sides up and embroider the center with peacocks, flowers, and trees in blues, reds, greens, and other colors, using *kantha* stitch. She knew that in the Bengali *kantha* tradition such quilts were embroidered with animals that were symbols of fertility and prosperity, which were wrapped around the new-born child for good luck, but said her designs no longer had specific meanings. *Kanthas,* patched cloths from Bengal and Bihar, were common in the nineteenth century and are one of the most interesting documented examples of recycling used saris and *dhotis* (Kramrisch 1989).[11] *Kanthas* were made by rural women according to their own designs, following general stylistic trends. Pieces of old white cotton *dhotis* and saris were layered; the typical Bengali colored borders would have their threads reused for the decorative embroidery. A lotus was placed in the middle, and the four cardinal directions indicated by trees. The rest of the field was decorated with animals, people, and both mythical and real contemporary objects, incorporating change within their form.[12] The whole quilt was then couched down in white threads that swirled around the colored representations, bringing a harmonious unity to the overall design and strengthening the fabric.

As well as being used to wrap babies, the quilts were used as seat covers and bedspreads and to wrap valuable or ritual items, and especially as dowry items for the nuptial bed. An expression of thrift, they eschewed urban sophistication, and were individual creations:

> The magic which underlies its purpose is that of love. . . . A *kantha* is given as a present, it is conceived with an outgoing mind and brings the entire personality of the maker to the person for whom it is made. . . . All of them are firmly stitched into a reconstituted, vibrant

wholeness. The *kantha* is a form, by textile means, of a creative process of integration within each woman who makes a *kantha*. (Kramrisch 1989, 83)

As quilts were often worked on by many women and became sentimental pieces and heirlooms, they joined family members together through production and inheritance. The very fabrics themselves brought together loved ones and joined them anew; the modest appearance of poverty deflected the evil eye while the designs provided protection.

The Making and Remaking of Persons

The reuse of clothing within the domestic sphere ensures that intimate pieces of cloth remain within the safety of the family. Favorite items of clothing tend to wear out more quickly than others through constant use, and several strategies are employed to prolong their lifespan. Their protective powers and emotional associations are saved from destruction by using and reusing pieces of cloth for children's clothing, by turning saris into suits, and by salvaging borders and *pallus* to use as decorative trimmings. Pieces of cloth are used as household furnishings, decorating domestic spaces with familiar swathes of a loved one's clothing that may be strung up as curtains over alcoves, made up into throws, or patchworked into cushion covers.

Whereas wearing hand-me-downs envelops the receiver in the unmediated form of the donor, imposing its own form of regulation, recycling transforms cloth through cutting and stitching. "Temporality is an intrinsic property of the object, which always exists in time, and will potentially signify the amount of time elapsed since it was created" (Miller 1987, 124). Recycling allows women to remake the spatial and temporal relationships inherent in old pieces of family cloth, mixing and matching origins and eras, creating wholeness in a new form: not just "making do" but making new. Enfolding the family brings together generations and collapses time, providing synchronicity and transcending the divisions between this world and the other.

It is the life histories of clothing that provide one of the strongest means of connecting earlier generations, or those who live far away, to the young and those at home. The sweater passed around the world from cousin to cousin and the sari handed down from a great-grandmother via several aunts continuously re-create their wearer's sense of self in relation to others. "A network of cloth can trace the connections of love across the boundaries of absence, of death, because cloth is able to carry absent body, memory, genealogy, as well as literal material value" (Stallybrass 1993, 45).[13]

Domestic Thrift

When reaching to wipe up a spilt dish or clasp a hot panhandle in the kitchen, women will often grab a piece of an old baby-grow (infant bodysuit or onesie), a boy's shirt, or half a skirt, conveniently hanging on a hook close by. Women in the middle of cooking may drape an extra half a sari around their waist or over their shoulder as an apron and generally useful rag. As clothing wears out, it is often cut up, replenishing the stock of rags always at hand for cleaning, polishing, and general use. Dusters and floor cloths have only recently entered the marketplace as new consumer goods, but are thought by many to be inferior to absorbent old cotton cloth.

One family was particularly resourceful in recycling old clothing. The grandmother, Kamala, was in her seventies, the widow of a judge; her family and her husband's were from Lahore, where she had been a school principal as a young woman. At Partition, they had left everything behind, sharing clothing and learning how to make do as the family established itself in India. She lived partly in the Progressive with her son (an advocate), his wife Bimala (whose parents had also come from Lahore at Partition), and their three teenage children, and partly with her daughter in Chandigarh. Kamala and her daughter-in-law were responsible for managing the family's old clothing. Kamala turned her old saris into *dupattas,* adding trimmings if she had them, or into handkerchiefs for the hot summers when "everyone is always wiping their faces." Her adult son's cotton shirts could be made up into smaller blouses, and saris could become slips, thin and light for summer. These clothes can be worn around the house, where nobody minds what one looks like. Her teenage granddaughter's clothing was brighter and more patterned—her long skirts became pillowcases, and parts of her suits were sewn into round cushion covers. Kamala and Bimala spent significant time and effort getting the most out of their clothing. Bimala's younger girls showed me a blanket of stripy knitted wool. They had unraveled two or three sweaters themselves and taken the wool to a shopkeeper in the Kingsway Camp, Old Delhi, to have it knitted up for Rs 150.

Thriftiness at home makes it easier to dress smartly at work, at school, and on social occasions, as is expected. The clothing chosen for young middle-class children and adolescents often embodies the tensions created by their family's simultaneous concerns with thrift and social status. Janaki stressed the cheapness of her clothes, which was a source of real pride for her, throughout our discussion about her wardrobe, and made negative comments about neighbors' expensive wardrobes. Janaki's brother had himself worn much of his father's clothing as a child, and her son Arun now wore his uncle's clothes, including a pair of Scottish tweed trousers purchased in the 1960s. These would have been expensive on an Indian family budget, especially given the poor exchange rates, and were a prized possession; in order to gain access to such clothing one had to have had friends and family members who could travel abroad, as at that time clothing imports were banned and Indian wool cloth and tailoring was of poor quality. But their good quality unfortunately ensured that the clothes did not (or would not) wear out even when they had become unfashionable.

Janaki had recently been to the U.S. to visit her now grown-up brother. The trip meant that Arun now not only sported the spoils of the last generation, but also had the most up-to-date new clothes and toys from the U.S. to extend his status as the boy to be envied, with tales of Disney World to authenticate his new Mickey Mouse watch. The tweeds and the watch earned his mother a reputation both for thrifty housekeeping and for providing her son with the most desirable consumer goods. Arun wore his tweed trousers for playing outside in winter; their very out-of-date strangeness perversely stood witness to his family's long-standing ability to bring high-quality clothing back from "outside," i.e., abroad.

Most children in the Progressive had relatives abroad, and therefore some access to foreign branded goods; but such things were often handed on from older cousins, and younger children, informed by satellite TV of the latest novelties, might not get them. Although such novelties remain the pinnacle of desirability for the children, their parents consider any American T-shirts and shirts, made of soft cotton, and foreign-branded jeans "the real thing." Again, the values of thriftiness and self-sufficiency must be balanced against those of newness, quality, rarity, and exoticism. As Sangeeta acknowledged with some irony, although she is pleased that her children are given good clothes by her sister in the U.S., she has ambiguous feelings about them: "My sister brings all her old clothes back to India, and some new ones too . . . all with brand names, and she goes through the prices of each one." But she continued, "In India you would never divulge if clothing was second-hand. My own daughter is brought up not to be at all embarrassed by hand-me-downs from the U.S. But *my* mother quietly tells her granddaughter not to advertise the fact—'there is no need to tell everyone.'"

For some children, being a younger child meant never having new clothes. Sangeeta was the youngest of three sisters and longed for something new for herself. In contrast, Usha was an only child who used to admire her elder cousins' new acquisitions and wait impatiently for the day when they would become hers. For the proud, thrifty middle-class family with middle-aged parents, the recycling of everyday clothes within the network of friends and family is a moral norm, but not one necessarily spoken about openly: older generations perceive it as rather shameful to admit that clothing is passed on, whilst the younger, upwardly mobile classes want to be seen to buy new. "Quality clothing should last"; yet these parents find their values at odds with both their elderly parents' values and the younger generation's.

Among the inhabitants of the Progressive, thrift had a high moral value. It helped to regenerate the extended family, but at the same time was associated in everyday parlance with the creation of the post-Independence secular nation-state. Middle-class families who described themselves as "socialist-minded," very much in the Gandhian tradition, made a point of linking thriftiness and the refusal to flaunt wealth in everyday life with the development of a more egalitarian society. However, being able to gift new clothes on religious festivals and dress up well for functions was equally important. Sangeeta described the different attitudes held by different generations of her family:

> *If you are a traditional Hindu then your kids should have*
> *new clothes for every festival. For example, Vasant Panchami*
> *is a festival to welcome spring. Everyone should have a new set*
> *of yellow clothes. When my grandmother was alive you did get*
> *them, but my mother was from a more Nationalist tradition of*
> *thrift and was not into waste. (. . . At my grandmother's house*
> *you were not finished eating until there was a pile you could not*
> *manage on your plate, but with my mother there was no wasting*
> *allowed. . . .) My father was into conspicuous consumption, but*

now my mother just dyes a corner of a few new white hankies
with temporary turmeric paste and sends them along to us as a
symbol—the color washes out and they can be useful . . . [but] it
does have to be new cloth.

The balance between appropriate conspicuous consumption and the avoidance of waste has become more difficult to achieve since the early 1990s, with more money in circulation and imported goods increasingly available. The ambiguities surrounding these concerns highlight the minutiae of struggle for social status and the moral high ground. The use and reuse of clothing is one of the most visible indicators of this struggle, and it is women who are both primarily responsible for thrifty household management and the vehicles for the family's vicarious consumption.

Spending on heavy silks and good-quality clothing can be offset by judiciously reusing and recycling cloth within the household. As Gudeman and Rivera point out in the Colombian context (1990), in the domestic economy good management ensures that money will remain after necessary outgoings; "making savings" and "being thrifty" are ways of skillfully controlling the relation between resources and expenditures. Thrifty behavior includes making leftovers into useful things, transforming remainders into expenditures, and wasting nothing. Crucially, it results in the hoarding of remainders as a reserve for the future: "Thrift, along with hard work, has long been identified as 'the bourgeois virtue.' Parsimony is said to be the practice of the rising and prosperous middle class, being a willingness to forgo present pleasures and amusements in the service of the future" (Gudeman and Rivera 1990, 171).

Thrift is a form of deferment; the value that is preserved is then devoted to a transcendent entity, typically the household. However, in his study of shopping in north London, Miller claims that "thrift has come to supplant the house itself as the process by which economic activity is used to create a moral framework for the construction of value" (Miller 1998, 137). The Indian consumption boom that began in the 1990s may well result in a change of attitudes, but thrift is currently still used to save for the house and family and to maintain their status. But in the cloth economy, what is "saved" may be the intimacy in a worn fabric, an auspicious influence, a memory, or the potential to enfold children. When clothing possesses none of these qualities, or even possesses qualities that its owners would rather be rid of, then the returns to be gained from different means of disposal can be calculated according to various registers of value. As I will show, such moral returns can include the knowledge that obligations to servants have been fulfilled, that a good deed has been done by handing clothing to the poor, and that something more useful has been received in return for the unwanted clothes and thus expenditure will be reduced. These dispositions form part of the modern *habitus* of middle-class urban women, whose practical mastery is the use of strategy in the art of daily life (Bourdieu 1977).

Negotiating Boundaries and Maintaining Status

So far, the giving and sharing of clothing has only been discussed within families. At the margins of one's networks of friends and relatives, relative statuses may be unclear and the permeability of domestic boundaries uncertain. This is especially apparent in discussions of children's clothes, which are frequently outgrown, should not be wasted, and therefore provide frequent opportunities for relationships to be renegotiated. Madhuri had only moved to the Progressive six months previously, although she already had several family friends and colleagues living there. Recently a neighbor had called around, asking whether Madhuri would mind if she offered her some leggings for Pia, her daughter. Madhuri was pleased: "I knew where they came from, so it was O.K." But such interactions can be fraught with difficulties. Sangeeta had once offered a neighbor whom she knew well some of her son's clothes, but felt that she must have insulted her greatly by doing so—relations had been somewhat strained since then, and certainly the neighbor's boy had never worn any of her son's clothing.

Purnima revealed with some embarrassment that her husband did not even like their children to wear hand-me-downs from cousins, and certainly never from neighbors—he did not want them to hear other children comment, "You're wearing my old T-shirt," as he thought it would make them "feel bad." He preferred to take his family to "branded" shops to supply them, as he considered it better to buy two good items than ten inferior ones. But his strategy, which was presented as a way of protecting his children's pride, neatly removed the family from all potentially fraught situations and set them apart from local exchange networks.

Cloth that binds families together also mediates between the home and the world. The exchange of clothing between families provides an opportunity to evaluate their relationship, defining "insiders" and "outsiders" and their relative positioning. Accepting children's clothing in good condition from a close friend or neighbor marks a tacit agreement that such clothing is not really "waste" at all, but more akin to the treasured party dresses passed between friends, coveted items worn by social actors not yet fully incorporated into social hierarchies. Such an exchange defines both parties as equals, as insiders within spheres of exchange. But old clothing is more often conceived of as *jutha,* polluted leftovers, which ought to be passed down the social hierarchy.

As the basis of orthodox Hindu beliefs regarding states of ritual purity and pollution resides in exchange between unequal persons, the acceptance of proffered "leftovers" is usually an acknowledgment of social subordination and dependency. This link between reception of cloth and subordination has well-documented historical antecedents and is not confined to Hindu belief systems. Referring to the Mughal imperial custom of giving court visitors *khilats,* or robes of honor, Cohn stresses the importance of understanding the substantial nature of authority in India to evaluating cloth exchanges, as they are a transactional medium.

> Clothes are not just body coverings and adornments, nor can they be understood only as metaphors of power and authority, nor as symbols; in many contexts, clothes literally *are* authority. The constitution of authoritative relationships, of rulership, of hierarchy in India cannot be reduced to the sociological construction of leaders and followers, patrons and clients, subordination and superordination alone. Authority is literally part of the body of those who possess it. It can be transferred from person to person through acts of incorporation. (Cohn 1989, 312–13; and see Gordon 1996)

The belief that harm can befall the recipient of another's clothing depends upon a belief in its transactional quality, which can have effects beyond those of ritual pollution or subordination. Gifts of clothing are ambiguous, and fear of them has led to a rich Indic folklore of "poisoned dress" stories or, concerning the Mughal era, "killer *khilat*" legends, in which the intentions of the *khilat* have been inverted.

> A *khilat* could be a loyalty test or a contest of wills. Myriad ambiguities
> in the custom could be exploited; one could manoeuvre a rival into
> accepting a *khilat* that hid hypocrisy, treachery, even poison. . . . The
> risk inherent in accepting a robe was compounded by the knowledge
> that cloth could actually carry contagion. . . . The practice of giving
> contaminated clothing to outsiders was a traditional folk ritual to
> cope with the fevers and epidemics that raged in Mughal times. . . .
> As a "second skin" that can protect the wearer, cloth can also literally
> endanger the body by flammability, poisons absorbed by the skin,
> and disease transmission. (Maskiell and Mayor 2001, 27)

Bayly quotes from the *Laws of Manu* 4:189, according to which receiving
an inappropriate gift could destroy a man's longevity, body, sight, energy,
or children, depending on its nature. A garment could destroy one's skin;
"cloth, then, was almost as integral to the person as his skin," and a ma-
liciously given garment might reduce a person to ashes (Bayly 1986, 291).

In contemporary middle-class Delhi, concerns about hierarchy and
status, dirt and pollution remain inherent in the disposal of used clothing;
in this context, caste can be seen as only one of many factors contributing
to status and identity. Both Béteille (1997) and Fuller (1997, 21) recognize
that caste ranking may be privately acknowledged or utilized as a cultural
idiom for other hierarchies such as class, while caste endogamy appears
to be maintained with some (as yet unquantified) strength. Residents of
the Progressive may come from all over the country and from a variety
of castes. Living in the city ensures a degree of anonymity for all but
close friends and neighbors, and daily routines resulting in contact with
a variety of people are balanced by a protected home life where certain
rules of purity and pollution may be upheld. Religious concerns are only
one of many factors influencing the interactions between the home and
the world. The housing society was a nascent community whose members
could not be certain they would be always treated as equals, therefore
the status of the cloth that was passed from person to person was also
contextually ambiguous; it was often, but not always, classified as *jutha*.

Giving to Servants

The difficulty of getting rid of old clothes is a major factor in the building
up of stocks of them, either in the wardrobe or separated out into piles of

worn-out cloth. The usual way of disposing of garments unwanted by one's family is to hand them on to servants and their families. The flow of used cloth downward is a feature of gifting cloth both to younger generations and to servants.

For the many upper-middle-class society residents who own rural land and maintain close links with their natal villages, this gifting is a form of traditional largesse to the servants and laborers whose families have been known to theirs for generations. As these urban families now travel home only for annual gatherings or important celebrations, women tend to stockpile clothing in suitcases in anticipation of the next visit. Such connections between urban and rural households are found across India, as far apart as Kerala, Bihar, Gujarat, and Assam. The distribution of old clothes across the subcontinent requires a great deal of arduous organizational effort, but is still thought by many to be the proper thing to do.

Most women also give clothing to their domestic servants in Delhi, to the daily maids, the *mali* (gardener), the *presswala* (ironing man) (figure 4), and the children of the contractors associated with the society. Rohini, like many women, keeps fifteen or so old, soft cotton suits to wear in summer. When she is bored with them and they are too worn out, she just gives them away, to her maids or to servants who ask for them. Many maids frequently ask for some clothing for themselves or their children, and such "gifts" are often regular and expected. More formalized gifting to servants is expected on certain annual festive occasions. At Holi, clothing is given to maids and to other workers in the society. The fact that such gifts are made largely to people who work directly for middle-class households reveals them to be part of a formalized gift-giving system. Women stressed that, although "customary," the gifts are not considered to be direct payment for specific services, are not openly valued or mutually agreed on as a remuneration (in contrast, see Raheja 1988). Most women also wanted to ensure that the maids and their families would not have to buy clothing, either new or from old-clothes dealers, and made an effort to find "someone who could really use it."

Yet most women are extremely careful about the quantity and quality of clothes they will pass on to their social inferiors and are particular about whom they give to, hoping and expecting that the recipient will indeed wear them. Women would sometimes refuse a servant's request, and many said they would only give a servant one or two items of clothing at a time, perhaps every three or four months, to ensure that they only had enough to cover their needs and they would have to wear them. Many maids were obviously mistrusted; women feared that they might misuse their gifts by selling them on to others, and in fact this was a significant conduit for the commodification of old clothes. Some women acknowledged their servants' right to make extra money this way, but other women couched their dislike of the possibility of resale differently; they perceived reselling the gifted clothes as cheating them, and resented that their servants might be better able than they were to maximize the monetary value of a garment.

Meghna thought carefully about the role of clothing in her relationship with her servants. She often gave cast-offs to her maid, whenever she had something to give, and—unusually—she reluctantly admitted that really, if it was her prerogative to give them away, it was then her servant's to sell them, that this was the "survival of the fittest." However, although she had herself bartered clothing away in the past, she no longer did so, feeling that the poor should not have to buy their second-hand clothes from the dealers but should be given them directly.

Meghna gave her maids new clothing, as distinguished from cast-offs, at Diwali, Dussehra, and Holi, but most importantly on the anniversaries of the death of her father and her husband's parents, in remembrance of them. Three or four times a year she would gift her maid with a woolen outfit, a blanket, a shawl, or something similar, and she had been doing so for the last fifteen years. She used to give clothing to an orphanage or school for the blind on these anniversaries, but decided that one ought to "nurture one's own plants," so she began to give instead to her own servants. For Meghna, doing so was a means of giving thanks to God and of expressing contentment with what she had, compared to others who were less fortunate. She considers these new clothes to be *daan*, religious gifts.

Saris and suits that are silk or richly decorated are rarely given to servants unless they are of very poor quality or extremely unfashionable, being thought inappropriate for their status, "too good for the maid." Women feel that if they have spent so much money on a garment, perhaps wearing it only a few times before it was torn or stained, it is a waste to give it to a servant. Servants who regularly work for the family should be reasonably turned out, with a pleasing appearance, reflecting the taste and status of the household. But underneath such sentiments lies the uncomfortable fact that one's maid should not appear as well dressed as oneself.[14] In fact, it is extremely difficult to give such fancy clothes away. If they are not recycled, they tend to accumulate in the wardrobe until a better solution comes along. But old cotton clothing and synthetics, which may have been owned by several family members already, can be passed on. A sari may be faded but its threads may still be strong, or the "seat" may have gone but the garment has some life in it yet.

Sometimes women use the services of each other's maids, especially live-in maids who may be spared for a couple of hours to help cook or clean for a party. As their habitual employer pays their wages, the women who borrow them will often give them an old suit or a scarf instead by way of recompense. Bulbul had a live-in maid, Anita, who often ran errands for Bulbul's older friends, receiving an array of old suits dating back to the 1970s in return, often worn-out party clothes in once-fashionable synthetics, which Anita would wear daily to do her chores. Other employees of the society might be given clothing if they had helped a family with a particular chore or had asked for something for a child at home. One informant used to leave very tatty old clothes on her doorstep with the rubbish, unwilling to offer them directly to the sweeper as a gift, but knowing he would probably make use of them.

Domestic servants occupy a rather ambiguous place in the network of personal relationships. They are at once social inferiors, substantially outside the kin networks of the family, and yet insiders, who may grow up and live within a family for years and are privy to its private concerns. Giving unwanted, "rubbish" clothing to them allows the giver to reinforce the social distance and hierarchy of the relationship; the gift acts as a prism through which relative statuses can be conceived. Yet the ambiguity of these relationships also allows benefaction and even affection to be expressed; givers can wrap servants in garments that perhaps once clothed their own children. Maids' acceptance and wearing of such items brings them into the family. These relationships are made visible through the gift, but their inequality is manifested through its material properties.

Such difficulties with the minutiae of giving highlight the wider ambiguities of the mistress-maid relationship. Adams and Dickey (2000) point out that the "domestic" is related to, but not contained by, the physical boundaries of the home; it is a site of production of societal power structures. The "domestic" refers not only to the space but also to the meaningful practices which take place there, leading to questions about membership in a household, to servants' being both "ours" and "not ours," and to uncertainty about what is private and what is not. The hegemony of the home is an instrumental, ambiguous hegemony, constructing inequality, domination, and resistance. Employers and workers interactively construct opposing identities out of their experiences with one another. "Such identities are necessarily fluid, positioned, and contingent. Like the relationships that give rise to them, they are also continuously negotiated and constructed in tandem through 'we-they' contrasts" (Adams and Dickey 2000, 2). These relationships need constant rejuvenation, fortification, and modification. In fact, research into the wardrobes of maids showed that almost in their entirety they are constructed from employer's gifts, yet these gifts are not incorporated in ways that their givers might have expected; maids have strategies for resisting the identities and value systems imposed upon them.

Donating to Charity

Giving clothing to charitable organizations or directly to the poor is problematic for many families, and a large proportion of informants admitted that they never did so. Popular discourse blames institutional corruption within charities, the failure of the state to provide for the poor, the unsuitability of the clothes available, and the possible unworthiness of the recipients for the lack of such donations. Individuals who did give clothes regularly to the poor always had a specific link to an institution that they trusted, such as a particular school or home. Charitable giving in India may also be directed to religious organizations such as Hindu ashrams or Mother Theresa's orphanage. The staff at the Sri Aurobindo Ashram, New Delhi, told me that most of the clothing they are given is packed up and sent south to their headquarters in Pondicherry for redistribution.

Natural disasters such as the cyclone that hit the coast of Orissa in autumn 1999 elicit calls for clothing, blankets, furnishings, tools, and equipment from national and state governments, NGOs, and religious groups, and many people give willingly and generously. For the Orissa appeal, the Progressive organized its own collection on behalf of a local agency, coordinated by three of the women residents, who received several large bags of clothing. But it is well known that the inefficiency and corruption of both the state and some NGOs hampers efforts to collect and distribute relief funds and goods of any description. Soon after the Orissa cyclone, media reports were appearing of truckloads of donated saris, suits, and *dhotis* left undistributed, and of middlemen selling them off to local dealers.

Even if such schemes were reliable, women often claim that their clothing is unsuitable for the climate in south India (or any other region), that styles are different, that their clothes might not fit those in need, and even, tellingly, that poor people could not wear such "good" clothes, with gold or silver *zari* borders or embroidery patches.[15] They express concern for the recipients: they would be embarrassed to look so shiny or fine. It is often claimed that the lower classes are reluctant to "step out of caste." Yet, as is evident in the fairly strict control over what type and quantity of clothing is given to household servants, middle-class women are equally reluctant to afford them the opportunity.

Except after occasional disasters, the role of organized charities in recycling old Indian clothes was limited in 1999–2000; I made a point of speaking to the directors of NGOs reportedly involved in clothing, but never found any systematic organization. When asked about organized charity, middle-class housewives often expressed a desire to help poorer people, but stated that in the suburbs they had no reliable means by which to donate things.[16] An NGO that ran a "materials mobilization team" did collect and redistribute small amounts of winter clothing but focused mainly on computer equipment and hardware, saying clothing was too difficult to collect and most middle-class people gave it to their servants. Another NGO ran a small shop where products it made from surplus material from the garment industry were sold in order to raise money, but although they were occasionally given old clothes, they felt their customers' middle-class sensibilities would be affronted if the clothes were sold in the shop (to other middle-class customers). The charity sold the garments on to old-clothes dealers who traded in poor areas and slums, rather than giving them away themselves. There are no charity-run outlets for second-hand clothing aimed at the middle-class bargain hunter or trend-setter as there are in the West, because clothing cannot simply travel up the social hierarchy.

Susan liked to stress that she only wears natural fibers. She gave away her cotton clothing, which wears out quickly, to her manservant, and, unusually, she admitted to me that he is expressly told that he must use it within his family, not pass it on, for fear of its being used for *tona,* black magic. Her husband John's clothes, old utensils, extra gifts, old coats, shoes, and so on were packed up in a huge carton and sent to Mother Theresa's orphanage in Delhi. She referred to "people's habit of getting something in return" for their old things as a detestable "Marwari culture," in which "the attitude is to not give away a pinch." She believed that the orphanage really uses his clothes and does not sell them on. All the people there are dependent on donations, and she got a blessing from the sister. After the blessing, she had "a good feeling" that came from having done the right thing. Susan was sure that Hindu notions of charity were different from Christian ones, which include the notion of the "deserving poor," the lack of emphasis on accumulating wealth for its own sake, and the idea that the donor is morally improved by helping someone in need.

Old clothes that are given to the faceless "poor" do not fit into the system of gifts analyzed by Parry (1994). Cast-offs do not fall into the category of *bhiksha,* gifts given to ascetics or religious beggars, nor can they be considered to be *daan,* religious gifts, because of their potentially polluting materiality. Neither do they form part of a formalized system of *neg,* gifts given in return for services, as there is no relationship between the giver and the receiver. Equally, such anonymous donation of cast-offs does not fit into the complex local exchanges such as Raheja describes in north Indian villages (1988). In fact, without any background knowledge of the poor, a giver can never be sure of the social status of the recipient.

Kaviraj has highlighted the morally ambiguous status of beggars in Indian society. Beggars can be destitute, fallen on hard times through hard luck, disease, or disability, or they can be religious renouncers who have chosen a higher path of living by begging. According to him, the lack of differentiation between these two types of poverty prevented beggars, especially those based in pilgrimage towns, "from being treated with undivided contempt" in traditional Indian society. "Householders consequently had a mixed moral attitude towards beggary; they felt they earned moral virtue by showing compassion to those less fortunate than themselves and respect for those who were more morally worthy" (Kaviraj 1997, 88, 89). But in the society he proposes, the community as a whole took care of its poor according to a common code, unwritten and undefined; this is not the case in a contemporary mega-city, or even in large pilgrimage centers such as Banaras.

The victims of the more systemic symptoms of poverty in urban Delhi, such as disease, malnutrition, violence, and the lack of shelter, health care, and education, appear to occupy an uneasy position in everyday discourse amongst the average middle classes. Delhi is not a major pilgrimage destination, but a vast conurbation growing exponentially through urban migration of the poorest sectors of society. The contemporary middle classes can continue to show their disdain for the poor, partly because theoretically various city and state bodies are supposed to provide for them (Varma 1998). This contempt is often mixed with a belief that many of the wandering ascetics, *sadhus,* they encounter in the city (alongside the many beggars who line up outside temples in the early evening) are charlatans. Many of my neighbors believed that most of them were cheats and liars. Few people, if any, are "deserving" of charity; if there is doubt, a couple of rupees can be given to a beggar outside a temple without great loss, but much clothing is too valuable to be wasted this way. So the poor without connections end up having to buy old clothing in the marketplace.

Rekha, the recently married woman from a wealthy family in south Delhi, said that she gave most of her (extremely expensive) seasonal wardrobe to her local ashram every couple of years in the hopes that the residents and staff could use them. I asked her whether she thought the religious and lay members would actually wear such gorgeous saris or would sell them on and use the profit for more basic necessities. She looked quite shocked by the idea; it was important to her that they utilized the clothing themselves, however inappropriate her rich silks might actually be as clothing for an impoverished urban migrant. In fact, her confident statements were likely to be intended to give an impression of wealth and largesse, as she had not been married long enough to have disposed of many clothes in this way. However, Banerjee and Miller (2003) have identified elite women who purge themselves of clothing on similar scales. A wealthy friend of Rekha's claimed to distribute her own clothing by a method she called "red-lighting." She would drive around the large ring roads in Delhi, and at traffic lights fish out T-shirts for the young boys who habitually sell dusters, cigarette lighters, and TV listings. She tried to find ones that would fit well before the lights changed and she had to drive off.

The ability of the very rich to give away valuable saris and clothing to the very poor can be linked to the deliberate destruction of value. Rekha's family was described as "Punjabi nouveau riche" by one of her friends, and her extravagant lifestyle, combined with her excessive display of wealth, suggests that she fit Bataille's model of functional expenditure. He claims that the agonistic model of expenditure found in the potlatch exchange has been replaced in the market economy by more stable forms of wealth, but that expenditure is still required to acquire or maintain rank. "In spite of these attenuations, ostentatious loss remains universally linked to wealth, as its ultimate function. More or less narrowly, social rank is linked to the possession of a fortune, but only on the condition that the fortune be partially sacrificed in unproductive social expenditures" (Bataille 1985, 123).[17]

The worst disaster to occur in India since my original fieldwork in 1999–2000 was the tsunami in December 2004. Over a year later, the website of Goonj, an NGO, echoed almost completely my findings from neighbors five years earlier. Warehouses in Chennai were overflowing, inundated with inappropriate clothing, which Goonj attributed to "issues of mis-match in terms of urban v/s rural attires, insensitivity, lack of awareness and callous attitude towards the dignity of victims." Although Goonj had initially been founded in 1998, no one I talked to in my early fieldwork mentioned it; now it has rapidly grown into a nationwide movement. The NGO aims to turn "a complete wastage after a disaster . . . into a resource for the poor" by sorting all the donated clothing and redistributing it across the country to the needy, while turning unwearable cloth into sanitary napkins for poor women (Goonj 2005).

Goonj is also trying to bridge the gap between rich and poor by promoting the concept of *vastradaan,* donation of clothes, to potential volunteers and donors, so that unwanted clothing can be channeled to the poor on a regular basis (Goonj n.d.). Distinguishing between development and charity, it suggests that while the distribution of clothing might constitute charity in Western countries, it does not in India. In India the question is one of survival for half the population, and the giving of clothing frees up meager aid resources for other developmental needs. "On a macro level, divide the society in two parts, one is Donor, the other is Receiver. Always remember that me and you as donors never buy it back, the buyer is someone living in a small locality, a rickshaw puller, a labourer, or a slum dweller." Interestingly, the web page describes how pleased people were to have the opportunity to get rid of mountains of clothing, as there hadn't been a disaster for some time. It appears that, with consumption rapidly growing amongst the middle classes, donations to the needy have indeed been hampered by the lack of infrastructure, yet Goonj found it necessary to stress development over charity as the principle underlying donations, tapping an alternative, acceptable ideology.

In 2007, the Resource Alliance and the Nand and Jeet Khemka Foundation named Goonj their Indian NGO of the Year and lauded the philosophy behind it.

> Without any financial burden to the giver or the beneficiary GOONJ has effectively used the wastage of urban India in different ways towards development activities in rural India. It has successfully transformed cloth giving from a charitable act into a developmental activity with its "Cloth for Work" programme. Its village level grassroots partners identify activities like road building, cleanliness drive etc., where the beneficiaries work towards the betterment of their own area and get clothes as remuneration for their work. (Resource Alliance and Khemka Foundation 2007, 4)

By switching focus from the religious to the secular register, the scheme has subtly reintroduced the notion of the deserving poor; by creating basic infrastructural work, it transforms the recipients of charity into laborers doing useful work and receiving payment in kind. In addition, the emphasis on "the betterment of their own area" associates the beneficiaries of middle-class largesse with both the creation of dirt and the subsequent cleaning up of their own unsanitary environments, reflecting a fairly common prejudice surrounding the deprivations of many of India's rural poor. The "Cloth for Work" program not only adds to the moral value of the donor but it is believed to bestow dignity on the receiver as well (5). Statements such as "Bihar's most backward 100 villages are utilizing material to address key developmental issues" (5) neatly sidestep any responsibility the state may have for these issues, whilst apparently imbuing the donated cloth with extraordinary political power.

Strategies of Closeness and Distance

Domestic conservation of clothing outwardly places a high value on continuity, and is certainly preferred for the most favored and treasured intimate possessions: the surrounding discourse suggests that the pre-servation of such clothing within the household is its most desirable fate.

Giving it away defines the boundaries of groups of family and friends, and is therefore fraught with difficulty; it nearly always results in clothing passing down the social scale as gifts that are not reciprocated and indeed must not be (see Parry 1986). These gifts are used to create ties of fictive kinship, establish hierarchy, and fulfill moral and religious obligations to the poor and unfortunate in the community. Benefaction and largesse in giving to servants and dependents is assumed to be the most common and habitual means of disposing of surplus clothing, one that properly utilizes the propensity of cloth to make manifest social relations and inequalities. Whether cloth is gifted to junior family members, friends, or servants, certain types of cloth are directed toward particular people, and the frequency and occasion of gifting may also be controlled.[18] The gift of a piece of used cloth is the gift of a detached fragment of oneself; it links the giver and the receiver, bringing them into a more intimate relation yet creating a hierarchy in the process. The negotiations involved in giving away used cloth are contingent on the struggle for closeness and distance.

Cloth is a valuable yet vulnerable resource; the destruction of old clothing in order to make new things enables the reordering of social relationships through the creation of new forms and the radical juxtaposition of old. What the everyday discourse conceals is the individual's initial act of riddance, which enables such social relationships to be remade; this is true of domestic reuse as well as extra-domestic recycling. Passing on clothing to servants also creates or reaffirms an unequal relationship. This creation or reaffirmation hints at the power of cloth to break and remake social relationships through the decay or destruction of its material form, a theme that will be taken up in the next chapter. The example of Priya's waistcoat revealed further options for getting rid of clothing through the marketplace, which she eventually took. This suggests that ridding oneself of connectivity in order to reconnect with the wider world is at least as important as preserving existing networks.

FIGURE 5. Mira and her sister Bina wait near The Progressive for customers to barter their old saris for *bartan*. PHOTO BY LUCY NORRIS

5 SACRIFICE AND EXCHANGE

The investigation of the market, both within India and in the West, reveals the complex trajectories of discarded Indian clothing, ejected from the wardrobe to be transformed into new coverings for bodies, objects, and places. This efflorescence of used cloth piled up in hitherto unimagined places raises further questions about how the value of clothing is constructed beyond the domestic realm. People often desperately need to divest themselves of unwanted garments, to be rid of cloth completely, along with its associations. There are several options for doing this: expensive clothing can be burnt or sold to collectors to become museum pieces, but more usually surplus clothing is bartered away for pots and pans.

Burning Cloth

Priya's waistcoat was worth too much to her to sell for Rs 50 to the *bartanwala,* who in turn did not value it as highly as a frying pan. The alternative Priya chose was to exchange it for the actual market value of the precious metal within it. The practice of burning *zari*-embellished cloth to extract the gold and silver has a long history, having been carried out by the royal courts at least since the Mughal period (Kumar 1999, 138). Precious metal thread would normally be incorporated into the silk cloth worn by royalty and the upper classes in a variety of ways, but in the course of wear and storage the silk would perish faster than the metal. Sometimes whole pieces of *zardozi* (raised *zari* embroidery) could be cut out and applied to a new garment (Kumar 1999, 112). Unless the worn garment had particular significance (if, for example, it carried auspiciousness or formed an inalienable part of a trousseau), it remained a material resource to be reused. The labor value of such textiles was insignificant compared to the value of the gold and silver within them. Just as jewelry sets were often given to the family goldsmith to be refashioned into modern styles, the value of luxury cloth often lay in the ability of its metal embellishment to be recycled. It was therefore relatively common for such elite clothing to be burnt to reclaim the metal within it, and selling it was a well-established option for the upper class, who had invested a good deal of money in valuable clothing.

Usha's mother had recently sold a Banarasi sari for its gold content. It had originally been Usha's grandmother's, and Usha had herself worn it a couple of times; it had considerable sentimental attachment, so she and her mother had hung on to it until it was really unusable. Usha had thought of trying to exchange it for kitchen utensils, but her mother was warned that the traders would cheat her and pay her for the silk, not the gold. (On a recent trip back to the Kinari Bazaar in Delhi, I did meet a group of Waghri women bringing saris and sari borders that they had managed to obtain through barter to the dealer for appraisal.) The sari had been of good quality but was not a historically significant piece, for which there is a very small market, and they "needed the money from it." Domestic objects containing valuable metals are sold at every level of society, and the practice ranges from thrifty behavior to distress selling.

Few saris owned by the middle classes have significant amounts of real *zari* work; the majority is artificial, or synthetic, *zari*. Many women claimed to be unable to tell whether the *zari* was real or not. I was told several differing methods for doing so, including by taste (real *zari* is tangy on the tongue), smell (it smells coppery if rubbed between the fingers), and sight (silver will blacken with age, and adjacent silk and cotton threads will degrade). But even if *zari* was known to be real, other crucial discontinuities of knowledge discourage recycling by selling it to metal dealers. Although some women knew they could get cash for their clothing in this way, most did not know of the existence of the precious metal dealers in Shahjahanabad and, when they heard of them, expressed surprise mixed with anxiety. Trips to the old walled city were potentially dangerous for women who did not know their way around and had no escort save an equally uncertain friend. Fonseca notes that the professional classes in New Delhi, in a characteristic sweeping generalization, describe the old town as a filthy, congested slum in which they can perceive no order (Fonseca 1976, 105), and indeed many of the women I knew had rarely, if ever, visited it.[1] Just as the Muslim artisans are isolated from the market within their workshops, so many of the middle-class women tend to remain in the new residential societies in Delhi suburbs; it is only the Waghri who move between these spheres.

Antiques

Whilst some recycled goods are flaunted as such, the origins of others are all but invisible to the eye; again, the concept of recycling in relation to cloth evades easy definition or classification. Thus the notions of "recycling" and value transformation blur the categorization of some textiles as retro saris, "antiques," and family heirlooms; the potentiality of the untailored sari, just a length of woven cloth structured through decorative patterning, facilitates this oscillation between value systems that apply to cloth draped on the body, to historical treasures, and to fetishized artwork.[2]

In elite circles in south Delhi, many confident women prided themselves on their historical and technical knowledge of textiles and the monetary worth of family clothing heirlooms. Textiles incorporating older designs no longer found or revealing techniques of construction now lost or dying out are especially valued. There is value in owning an original piece, and some are restored, mounted, and displayed in the homes of the wealthy.

Networks of petty traders still traverse the routes between villages and towns, quietly buying up pieces from respectable families who do not wish to sell their possessions on the open market. These traders sell them on to dealers, who offer pieces to museums and institutions, private collectors, and designers desiring original source material. According to staff at the National Institute of Fashion Technology, who were putting together a collection for their students in 2000, small jackets can sell for Rs 15,000, small tents and canopies for Rs 50,000, and larger pieces for up to Rs 100,000. The largest dealers, such as Bharany's in Sunder Nagar, employ skilled restorers and deal directly with Indian aristocrats and collectors world-wide. Conversations with curators at the National Museum revealed that collections of older textiles are built up through relationships with traders who offer them pieces as they become available, as are the collections of the Crafts Museum.

Not all antique textiles are sold as collectors' objects. One aristocratic woman rather snidely claimed that the up-and-coming "nouveau riche" bought up old textiles in an attempt to pass them off as their own heirlooms. The widespread dislike of wearing other people's clothing, for fear of pollution or more generally of bad luck, stops many from buying textiles to wear. One woman said that she would always wonder what had happened to the previous owner in order to make her sell it in the first place, suggesting that the inauspiciousness of distress selling clings to a textile. Older textiles made from natural fabrics are unlikely to be in good enough condition to wear. Sometimes women buy an original and take it to master weavers to have exact copies made; its worth as a pattern is an important component of the older clothing's contemporary value.

One interesting case study highlights the different calculations of value among the generations of one elite family. Shobha Deepak Singh came to my attention via a feature in the *Hindustan Times* (February 7, 2000) on her growing collection of antique saris: "Never mind if they're torn or stained, the 56-year-old sari collector feels they're more precious than antique jewellery"—a reversal of the commonly accepted value system which privileges gold as inheritance. She had been amassing an assortment of antiques, her own creations, and some new pieces copied from heirloom designs, and Vastra Shobha, her first large-scale exhibition, was held in the basement of her luxurious home in Sardar Patel Marg. Although she had collected saris for years, it was her only daughter's engagement six years earlier that had really inspired her. To put together a trousseau of a hundred custom-made saris, she had had to commission four of each design, because the weavers set up their looms for four saris at a time. She therefore bought three extra for every one she had commissioned, and sold them on to enthusiastic friends also hunting down trousseau items.

She also started buying old saris and having them restored, especially Banarasi brocades from royal families, Gujarati and Rajasthani prints, tie-dye, and *zari* work. Slowly, weavers and dealers started to contact her and offer her rare pieces. She clearly had a flair for design and strong business acumen; her designs were exclusive, often unique, having been based on rare originals, and she could adapt and modernize techniques to produce new combinations for the Delhi elite. In the exhibition, new saris were suspended from high rails and artfully backlit, while the older pieces were folded over hangers, interlined with tissue. A coffee table was weighed down with expensive, authoritative tomes on textiles and saris, such as the series issued by the Calico Museum in Ahmedabad and international auction catalogues. Ostensibly a resource for interested clients and visiting students, they added gravity and value to the clothing displayed, declaring it to be of equal worth to those depicted in the printed texts.

Although she intended to sell saris at her exhibition, she was reluctant to get rid of certain pieces that she had had for years, and took me upstairs to a large walk-in wardrobe overflowing with more rare textiles. There were no established boundaries between her personal collection and what was fast becoming a successful business downstairs. She herself had received the Padma Shri, a presidential award, in 1999 for her work as director of the Shri Ram Bharatiya Kala Kendra (a cultural center for classical music and dance), and had wanted to wear a favorite old sari of her mother's that was in an unusual design and color combination to the award ceremony. Unfortunately it was no longer wearable, so she had managed to have a copy made. She felt the Delhi elite wanted a similar combination of exclusivity, aesthetics, and nostalgia in their saris, and wanted to sell only to those who appreciated fine textiles and could afford to buy them. New saris with real *zari* work cost more than Rs 20,000, she said, but certain people just spent Rs 5,000 on artificial *zari* and got rid of the sari after a few years. She set duplicates and highly unusual pieces aside; she wanted to eventually establish a museum and leave all her saris to it.

Shobha was realistic about the likely fate of her own clothes, knowing her daughter might not want them. She was establishing a new way of getting rid of them: turning them into antiques, literally priceless through being confined to a museum. She sees herself in the vanguard of establishing antique textiles as an increasingly acceptable form of stored wealth, following on from the growth in interest among collectors. Her daughter was living abroad and rarely wore the saris that had been so carefully picked out for her; indeed, most of them remained in Delhi to be worn on her occasional visits. Although commissioning the trousseau had given Shobha so much pleasure, once it was displayed and handed over, its function was complete. A prominent London dealer in high-quality Indian textiles commented that he is often approached by elite young women who are already trying to sell their trousseau before most of the saris have ever been worn.

Most of the families living in the Progressive and neighboring housing societies could not afford to buy the most expensive and exclusive textiles being created in India today. However, some women had one or two heirloom pieces in their wardrobes, perhaps Banarasi saris or fine pashmina shawls. These women were often from families that had had high status and wealth a generation or two earlier but which, for various reasons, were now living more modestly. Occasionally such pieces might be wearable, but usually they were merely kept in trunks and cupboards and handed down. Some women expressed an interest in preserving their wedding saris, believing that their value would increase as older designs became increasingly rare.

But if circumstances required (typically in cases of financial distress), items of high enough quality could be sold in the thriving market for antique textiles in Delhi. As Parry acknowledges, there is no conception of money being the "root of all evil" in Hinduism; indeed the accumulation of money can, in the right circumstances, be the fulfillment of one's *karma* (Parry 1989, 78). Women might approach dealers themselves if they had an idea of the value of their goods or had been prompted by hints from others that good deals could be struck. But their scant knowledge of the antique market usually made them doubt their ability to sell personal possessions. Charu, a middle-aged Gujarati woman living with her husband and younger daughter in Sainik Farms (an expensive development in south Delhi), came from a royal family and had married into one. She regretted having sold pieces of Gujarati embroidery and silk sari borders cheaply to itinerant traders a few decades earlier, primarily because they had increased in value since—rich women had not previously considered them to be worth anything.[3]

Bartering for Pots

Although it has probably always been difficult to redistribute excess clothing effectively, people seem to feel that the excess is increasing, that clothes are "more of a problem" nowadays. They suggest three main reasons for this. Firstly, in a new housing community in Delhi such as the Progressive, it is harder to hand clothes on to extended kin networks and residents of natal villages, their traditional recipients. Secondly, commercial products to furnish the home and use as cleaning cloths are increasingly available and have somewhat superseded the reuse of cloth within the home. Thirdly, the increasing speed of changes in clothing fashion may have a determining role—many women are acquiring more clothes to keep up with the changes. When these factors are combined with the move to smaller family units in apartment buildings, however, women begin to feel that they cannot keep outdated clothes that they would no longer wear but that are not worn out, a feeling that conflicts with the practice of thrift within domestic and personal spatial networks. The truly poor wear their clothing until it wears out, and there is never a surplus. But among the burgeoning middle class experiencing the consumer boom that began when the economy was opened up in the early 1990s, the traditional strategy of the rural gentry, petty urban aristocracy, and educated elite—handing on clothes to their servants and inferiors—cannot seem to keep pace with the accumulation of surplus high-value goods. A further option is increasingly becoming available to middle-class Indians: that of exchanging old clothing for kitchen utensils, *bartan*.[4]

During the day, as residents of the Progressive walk up the road to the local market, past the fruit and vegetable sellers' carts, they are likely to pass the *bartanwale* sitting on the footpath under a tree. Sometimes one woman sits alone, and sometimes she is accompanied by her daughters and daughters-in-law, but each family has its particular patch (figure 5). On the road a few feet in front of them is a dazzling display of shining *bartan*: stainless steel plates, beakers, small bowls, storage jars, cooking pots and pans, and perhaps a small rack. In addition there may be glass tableware, such as the newly fashionable molded-glass serving bowls with matching dishes, sparkling in the sunlight. To one side a range of plastic bowls and tubs in bright colors, used for household storage and cleaning, complete the tempting tableau.

The women arrive mid-morning with baskets full of *bartan* carefully balanced on their heads. After a day's work, they leave again, with bundles of old clothes tied up in pieces of old sari on their heads, replacing some of the pots they arrived with. During the day they make no attempt to call out and encourage women to stop, but allow the glinting surfaces of their

wares to attract attention. A potential customer always begins by determining the quality of the merchandise; Indian housewives pride themselves on their ability to judge the weight of the steel and how close-fitting
the lid is. The *bartanwale* then ask what items the woman has at home to
dispose of. Primarily they trade for clothing, although some old electrical
items, shoes, wristwatches, and other goods are increasingly sought after.

Once the quality and desirability of the old things are established in
principle, a tentative ratio is established, for example, "six pieces [i.e., six
cotton *dhotis*] for each metal beaker," with the understanding that the real
bargaining will commence when the clothes are viewed. If a particularly
desired item is not on display, the dealer can arrange to bring one later.
Once the woman's interest is aroused, perhaps then and there or perhaps
later on after a perusal of her family's wardrobes, she may arrange for the
bartanwale to visit her home. She may have her heart set on a particular
piece, such as a new bowl, or she may have an excess of clothing piling up;
whatever the primary trigger, the shining display is a means of luring her
old clothes out of their marginal spaces and into the marketplace.

The *bartanwale* are Waghri people from Gujarat and have an extremely poor reputation.[5] During my initial investigation into the system
I was repeatedly told by middle-class acquaintances that the Waghri were
dangerous people, untrustworthy, and that allowing them into one's home
was unthinkable—the women were known to be merely "casing the joint"
for a return visit by their husbands, whom my informants branded thieves
and criminals. Yet every day they would depart from the local market with
bulging bundles of cloth on their heads, the result of the successful pris-
ing of clothes from wardrobes. The sheer volume of second-hand clothing
found in the markets, from silk saris to old T-shirts, made clear that unwanted garments were covertly discarded via the *bartanwale* on a regular
and massive scale.

Many of the residents of the Progressive claimed not to barter their
clothing for pots. Some said that they felt it unnecessary to get a "return"
for their old clothing and that they saw the exchange as demeaning. As
the previous chapter described, they kept and reused material within the
family, and had extended networks or links to specific charitable organizations for handing on cast-offs. Often people appeared rather horrified
at the idea of bargaining with clothes, especially women who considered
themselves to be upper-middle-class. It was important to them that they

look like people who could and did simply give surplus away as a moral obligation to dependents. However, it was clear from the huge bundles of old clothes that the local area itself was a ready source of trade for the *bartanwale,* and several women, when urged to be candid, considered trading clothes to be quite a good way of getting "something for nothing."

A few residents of the Progressive did admit that they, or their mothers-in-law, had traded with the *bartanwale* "in the past" or "two or three times." Purnima's mother-in-law used to dispose of their joint family's cast-offs this way, but had stopped two or three years before. She no longer liked to barter, the stainless steel was inferior, and the dealers kept cribbing about the quality of the clothes. They began demanding sari borders, which Purnima said she would rather use up herself and sew on a suit. Usha had also obtained utensils in return for some of her old heavy saris that had worn out over the thirty years since her marriage, but no longer had any left. The process of bartering forced the women to evaluate their old clothing in a new and often unwelcome light; the dealers might dismiss favorite old clothes out of hand, while sari borders appeared more valuable.

Sushila, a Brahmin Chaturvedi who grew up in Kolkata and had married within her caste, lived in the Progressive in a very large extended Brahmin family that included herself, her husband, their two sons, her husband's mother, his brother and sister-in-law, and their two children. She described the family as "not so modern, not backward, but medium," a culture where one respects one's elders. Her mother-in-law ran a thrifty household, and they tried never to waste anything—"if your work is done in a certain way you will save money." If they needed a bucket or a *bartan,* they would sort out their *phataa,* old saris that were unwearable or ones they were bored with, and pool them for bartering. However, it could often be a traumatic experience. Sushila herself enjoyed bargaining, but she could not face the Waghri women; only her mother-in-law had the skill "to talk smartly" with the *bartanwale.*

Women have to decide how to balance the moral expectations surrounding them: whether to keep valuable or auspicious clothes circulating within the family, hand serviceable ones down to servants, or thriftily exchange them for something more desirable. Their decision is affected by both their family's circumstances and the meanings inherent in every piece of cloth. The surplus clothes unsuitable for family use are sacrificed, and in the process the "remains" acquire exchange value; barter is the most appropriate form of commodity exchange by which to get rid of them, and in north India pots are particularly suitable to be exchanged for old clothes.

The Doorstep Exchange

Rohini was an assertive woman and volunteered to bargain some of her family's clothing, along with a box full of a mutual friend's unwanted garments, so that I could watch the process up close. At this early stage I had no knowledge of the market in second-hand clothing and what different articles might be worth. She had used the *bartanwale* before, to dispose of old silk saris that were "too good for the maids," but had felt cheated and exhausted afterward. "Once the exchange begins, they keep demanding more and more things, until you end up giving away things you wanted to keep, all for the sake of one or two *bartan* which you could have bought in the shop."

As we walked up the road to the *bartanwale,* her main advice was "Do not tell them which one you want; point to the biggest thing and ask what they want for it, then bargain very hard." She acknowledged that they had the advantage, as "they know you do not want the clothes, but you could get something nice for the kitchen, rather than fritter the household money away."

That day, the young *bartanwala* Mira was there with her sister Bina. As we looked through their display, they talked generally of value in terms of items of clothing: twenty pieces for a large steel container, fifteen for a medium one, six for each *thali* (plate), eight for a set of six small *katore* (eating bowls). We selected a set of six plates and bowls and a large container as potential acquisitions, and set off with Bina to Rohini's flat. Initially the guard was unwilling to even let her in; eventually he acquiesced, but demanded that she be escorted off the premises, unlike any other *kabadiwala.* She sat down on the doorstep of the flat and waited as we retrieved some clothing from inside. We had piled it up ready in the storeroom next to the front door, out of sight; the whole barter took place on the threshold, Bina sitting cross-legged, peering in through the door.

The number of items required for the *bartan* we had chosen had already fallen on the walk back to the flat, and we had decided to go for the set of plates and bowls, which now would cost thirty-six items in total.

Rohini began by bringing out the clothing she suspected to be the worst. The box of donated clothes contained many Western garments, since our friend had spent time living in Europe—we produced a selection of women's scarves (head squares), casual tops, swimwear, European-style underwear, and thermal underwear. These were all disdainfully refused. The clothes were to be worn by poor people—what would they do with these? Why would they buy them? She took a couple of embroidered women's shirts that could be worn by girls, and three rather matronly tweed knee-length skirts. She was equally uninterested in men's shorts, but eagerly took men's shirts and formal trousers; however, even though they were Levi's and Wranglers, she claimed that the labels were of no consequence. Blue and black were desirable colors; khaki pants and chinos were poor seconds.

At this point Rohini began to extol the virtues of the clothing, pointing out the high-quality fabric, the smart designs, the fact that the clothing was all European in origin and had belonged to someone in the diplomatic service. But Bina was unimpressed, and soon started calling out for Indian clothing: saris, *dhotis,* and suits. Rohini produced a pile of old cotton suits, and Bina started counting through them without checking their condition at all. Each suit, consisting of trousers (*salwar* or *churidar*), tunic (*kurta*), and scarf (*dupatta*), was counted as one item. Because the *dupattas* did not match the other pieces, Rohini maintained strenuously that each suit should count as two items; she eventually won the argument by again pointing out their good condition. Immediately Bina called for synthetic suits, much more desirable in the marketplace. Rohini called attention to the labels in some of the suits, which came from good shops in Connaught Place and south Delhi, but these were brushed aside. As the pile grew, it became clear Bina was miscounting, which triggered raised voices, another argument, and several recounts by Rohini.

Bina then split the pile of clothing in two, brought the container back into the bargaining, and upped her call for things, now asking for men's shoes and wristwatches or old saris, so that one pile of goods would be equal in value to the plates and bowls, the other to the container. In attempting to obtain more goods of better quality, she now allowed such things to be worth more: a pair of good men's shoes counted as an item each, a sari with a *zari* border might be worth five or six lesser items, a watch even more. Her bargaining strategy was to keep the value of items as low as possible, until the prospect of really profitable garments demanded that she give a little encouragement to winkle them out of us. At any point when the arguing seemed overwhelming she would produce the *bartan,* extolling its worth.

Eventually it was clear that we had no more goodies left inside, although Rohini had been persuaded to go and look through her husband's and son's wardrobes to see if they had any worn-out shoes. Rohini was fed up and feeling miserable—after an hour, it was time to wind up the bargaining. We could not offer enough clothes for the container, so the piles were amalgamated; however, Bina then started rejecting items again, in an attempt to obtain more jeans and higher-quality clothes. Eventually, the set of *thali* and *kattori* was handed over. Their total value in a shop would have been approximately Rs 200 to 250, and in exchange for them Rohini had bartered around thirty items, whose original prices were between Rs 100 and 1000 each.[6] Almost to add insult to injury, it seemed to Rohini, Bina asked for half an old sari to bundle the clothes up in, and with a broad smile departed.

Riddance and Renewal

The process of detaching from one's clothing is inherently ambiguous, at times encompassing both the pain of separation and the relief of being rid of unnecessary belongings and experiencing a sense of renewal. Clothes that are kept, cherished as souvenirs and mementoes, are nonetheless eventually given away, usually to loved ones in order to preserve their sociality. This understanding of conservation fits with Weiner's notion of "keeping-while-giving" (Weiner 1992). However, the barter of quantities of old clothing reflects women's desire to be rid of both the material and its sociality through strategies of exchange and alienation; that this destruction of relational value may be considered socially unacceptable explains their reluctance to admit to bartering themselves. But the ontological project of "being and becoming" underpins all these practices of divestment and resonates throughout the ethnographic data.

Baudrillard identifies objects as helping collectors to establish dominion over time by inserting the objects into mental sets. By classifying and arranging objects, people can interrupt the continuous flow of time, dividing it up in order to "resolve the potential threat of time's inexorable continuity, and evade the implacable singularity of events" (Baudrillard 1994, 14). Real time is displaced into the dimensions of a system. But in order for the living person to construct a future for himself or herself, the wardrobe (as a collection) must be constantly reconnected to the flow of life; new clothes must be acquired, and others must be cast out. The wardrobe can articulate the relationship between the visible world as it exists at one point in time and the invisible world of future potential. This potential has been explored by Pomian, who analyzes funeral objects, museum collections, gifts and booty, relics and sacred objects, and royal treasure as mediators between the visible and invisible realms, offerings to the transcendental realm. He conceives of the invisible realm as opposed to the visible in various forms: for example, man and god, this world and the next, the secular and the sacred, present and future. The invisible is always spatially and temporally distant; it has "a corporeity or materiality other than that of the elements of the visible world" (Pomian 1990, 173).

The wardrobe as a collection exists between the past-in-the-present and the future. It needs to be acted upon, to reconceive and make possible the project of the future. Clothing forms the cultural boundaries of the body, and is vulnerable to obsolescence and decay as people grow older, rendering the body boundaries unstable. Getting rid of it, as the body rids itself of exuviae, returns the body to a bounded state and simultaneously reinforces its social integrity.[7] If clothing is perceived as a representational image of a person (Hollander 1993) and as a detachable part of the body, then the "untying" of the binding relationship between the person and his or her unwanted cloth images can be understood as an iconoclastic sacrificial practice (Loisy 1920; Küchler 1997).[8]

As transient representations, clothes have their own decomposition built into them, and upon the centrality of their materiality rests the efficacy of the sacrifice. The icon must act as a symbol of both the sacrificer and the outcome he or she wants to achieve. Valeri notes that the destruction of the icon avoids the sacrificer's own death by representing it instead, allowing the sacrificer to carry on living, growing, and changing. Sacrificial death and destruction are also images; they represent the passage from the visible to the invisible and thereby make it possible to conceive of the transformations the sacrifice is supposed to produce (Valeri 1985, 69).

Both discarded items and things that are set aside, "discarded-through-treasuring" (e.g., an art object in a museum), are separated out from daily life and are thereby "made sacred" (Pellizi 1995). During sacrifice the sociality in the object is destroyed. This sacrifice translates thrift (saving, hoarding) into expenditure or destruction, and transforms what might have been ordinary acts of expenditure into the means by which the transcendent is constituted and affirmed (Miller 1998), here understood as the future. Indian women purge their wardrobes of undesirable clothing in order to create a new ideal state of cleanliness, auspiciousness, integrity, and connectedness, both for their own person and for their family, which is represented in the fabric of the garments themselves. Riddance of old clothing restores the integrity of both the body and the wardrobe, and helps to maintain the domestic household in a proper state.

Keeping the home clean and auspicious is a central part of women's work, and as Chakrabarty notes, "housekeeping is also meant to express the auspicious qualities of the mistress of the household, her Laxmi-like nature that protects the lineage into which she has married" (1991, 20). Quoting Raheja (1988, 43), he continues, "The negative qualities and substances that may afflict persons, families, houses and villages are seldom 'one's own,' they achieve their 'entry' though lapses in the performance of auspicious actions." Chakrabarty concludes that "auspicious acts protect the habitat, the inside, from undue exposure to the malevolence of the outside. They are the cultural performance through which this everyday 'inside' is both produced and enclosed. The household rubbish marks the boundary of this enclosure" (Chakrabarty 1991, 20).[9] Here it follows that it is in the acts of shedding clothing that the boundaries of the individual, wardrobe, and household are continuously re-created and reinforced, and in the process the contingent relationships between family members, servants, and traders are reconfigured at each transaction; garments do not cross preexisting boundaries of purity and pollution but create those boundaries as they are exchanged through mundane daily transactions (see Douglas 1966).

The wardrobe is an intellectual economy (Harrison 1995); when getting rid of garments, women use their knowledge strategically to decide between systems of disposal. Where old clothes are routed out of the family and household, they can be understood as remains, stripped of their sociality; barter allows women to create "something from nothing." However, clothing is "sticky"; it clings to the person to which it is attached, and requires something aesthetically slippery, ideally the shiny metal pots obtained in exchange, to smooth the path of divestment.

The Efficacy of Barter and the Lure of Pots

Contrasting reproductive gift exchange to the household service economy, Gell asserts that "'exchange' provides an escape route from a social order in which objects are transferred, and services performed, out of moral obligation, substituting for it one in which transfers and services can be conceptualized in terms of the schema of mutually advantageous exchange of sacrifices" (1992a, 152). In gift exchange, the objects themselves *are* alienated, but it is the identity of the donor that is inalienable and still attaches itself to the object after it has been given away (145). He concurs with Gregory (1992) that commodity exchange is defined as the "exchange of alienable objects between transactors in a state of mutual independence, and the exchange as one which establishes a qualitative relationship between exchange objects. When a . . . commodity swap has occurred, . . . the transactors involved are 'quits' with respect to that transaction" (Gell 1992a, 144). But Gell argues that gift exchange is actually modeled upon commodity barter, and it is only the social context of the transaction that distinguishes between gifts and commodities.

Exchange has certain moralities associated with it; "embedded" exchange is socially rich, but disembedded exchange such as barter is antisocial. Appadurai defines barter as "the exchange of objects for one another *without* reference to money and *with* maximum feasible reduction of social, cultural, political, or personal transaction costs"; it is the exchange of things with neither the constraints of sociality nor the complications of money (Appadurai 1986a, 9–10). The *bartanwale* are analogous to Simmel's stranger (Simmel 1971), who is both far from the community yet near by virtue of his presence. The trader's relationship to other members of the society depends on the nature of the exchange. Those with whom one barters should embody all that is "anti-social" about the process, so the outcaste itinerant trader is the epitome of the exchange itself. The Waghri use their bargaining skills to their advantage and bring their particular knowledge of the second-hand clothing commodity market to the exchange, enabling them to make a good living out of someone else's cast-off remains. But why are pots a suitable exchange item, and is there any significance to the pairing of these items?

Itinerant merchants trading brass and copper pots have long been a feature of Indian rural life where small villages did not support a resident craftsman (Miller 1985, 23). Tarlo has described the custom of trading pieces of peasant trousseau embroidery for pots in Gujarat, where it was begun by entrepreneurial *pheriya* (wandering traders) in the 1950s (Tarlo 1996b). Interest in such embroideries had been sparked in Mumbai after they were used for a costume in a Raj Kumar film. At that time in rural Kutch, peasant women's embroidery had no commodity value in the local market. Clothing styles were specific to each caste and region and were becoming old-fashioned. But when clothing was sold in Mumbai to a new market of Indian artists and foreign collectors, beyond the museum world's *cognoscenti*, it began to acquire a high value. "Obtaining stocks was easy because women would part with their embroidery for petty exchanges of *bidis* (local cigarettes), grain or a new kitchen utensil. . . . The exchange of embroidery for stainless steel began as an exchange of exotica" (Tarlo 1996b, 8, 12).

Tarlo's research suggests that the most important motivation for women to sell their embroidery was changing local sartorial styles, and that elite peasant families would exchange large quantities of trousseau items for a single steel plate back in the 1960s, although widespread selling also occurred in times of famine. New embroidery styles were favored instead, coupled with the increasing fashion for wearing mill-made saris. Tarlo mentions the attractiveness of the steel utensils as a factor; people wanted to replace their brass pots, which were laborious to clean, with shiny stainless steel, nicknamed "German Silver."[10]

The custom of exchanging clothing for pots was developed in Gujarat, and its spread to other major cities in north and central India can be attributed largely to the expansion of a few families based in Ahmedabad (Tarlo 1997). Its scale is now vast; tens of thousands of Waghri earn a regular living trading in used clothing, and Waghri women suggested that its exponential growth was at least partly linked to the collapse of the mills in Ahmedabad and the subsequent high unemployment amongst their menfolk. Ranges of molded glassware are now offered, while the dwindling supply of regional embroidery has been overtaken in popularity by decorated silk saris. How could a regionally specific trade be adapted so successfully across north India?

In Gujarat, peasant embroideries are decorated with mirrorwork, metal thread, and sequins to enhance their shininess. Inside small, round thatched huts in Kutch, "built-in" mud cupboards are decorated with designs echoing those found on the clothing, with inset mirrors glowing by lamplight. Displays of metal vessels, *mand,* are also common; a few brass pots are still seen, but more usual nowadays are multiple shelves crammed with sets of stainless steel *bartan.* The vessels reflect and augment the dim lights all over the room in ways reminiscent of a glittering *shishmahal* or mirror palace (a room found in royal residences across the desert states of Gujarat and Rajasthan). Today, in the cities of Ahmedabad and Delhi, Waghri households continue to display metal utensils, which are among the few items of material culture they possess, and any brass and copper vessels are cleaned and polished for festivals such as Diwali (plate 5).

Jain claims that the Gujaratis have a unique love for *bartan* (Jain and Patel 1980). Spare money is often invested in utensils; they mark status and are symbols of family wealth. Collections of *bartan,* including brass and copper utensils, grow over time and are given as dowries. Indeed, the major items in a dowry would consist of a trousseau of embroidered clothing and furnishings and a collection of utensils, together with a few pieces of household furniture.[11] Jain also notes that *bartan* was customarily given at ceremonies for a child's birth, at the ceremony of tying the sacred thread, at weddings, and on commemorative occasions after death. Utensils were distributed as gifts to caste members and sometimes gifted when returning from a pilgrimage—details of the journey were engraved upon the pot.

However, the appreciation of gleaming pots is not confined to Gujarat. Writing in the early twentieth century, Coomaraswamy claimed that "domestic brass is the glory of the Hindu kitchen, . . . cleaned daily and polished to a degree that must be seen to be believed" (Coomaraswamy 1913, 142). Today in the Progressive, most women have a good collection of steel plates, pots, and lidded containers that they use daily, and pots are often artfully arranged in sets and in size order on open kitchen shelving. Every shape of pot is available in gradated sizes, so that each set, once complete, can be laid out from small to large along the shelf, or stacked up from large to small in a tower.

The first day of Diwali is known as Dhanteras (*dhan* means "wealth") and is believed to be auspicious. After the spring cleaning of the house (and maybe turning out the old clothes), women may buy gold or silver coins, jewelry, and one or two new utensils before performing *puja,* worship, to Laxshmi, the goddess of wealth. At Dhanteras the local market near the Progressive was full of itinerant *bartan* sellers on the street, and the regular shops were doing a brisk trade. When asked, many middle-class women would say that a new *bartan* was a pleasurable acquisition, one or two adding, with a smile, that the kitchen always needed a new pot to add to the collection, much as they always needed a new sari. Pots are therefore an appropriate form of wealth for women, beyond their utility value, where the household's resources are converted from cloth to metal (much as the Muria Gonds in Gell's account [1986] populated the landscape with empty concrete houses).

That women desired pots and acquired them for themselves was also recognized by men, sometimes with an air of amused resignation. Domestic pots, both clay and metal, are part of the woman's realm, the household economy managed by the wife and mother-in-law, and are a source of pride and a means of aesthetic display.[12] At the upper levels of society, handcrafted pots have gained popularity as household *objets d'art* much as fine textiles are. In addition to their aesthetic appeal and their importance in women's dowries and the domestic economy, pots resonate with ritual significance and cosmological symbolism.[13]

Pots replace a disintegrating image of the relational self with a more durable aesthetic representation that is usually paired with cloth in dowries; value has passed from the register of cloth to that of metal pots, which are also used to cook and prepare food for the family in the most sacred space in the home. Women manage the wardrobes of the household and barter the whole family's clothing on the doorstep. Raheja has noted the spatial aspects of ritual: "thresholds and village boundaries are very often the points at which negative substances and qualities enter or are transferred from the house or village, and there are many actions and prestations that restrict the entrance and facilitate the exit of inauspiciousness at these places" (Raheja 1988, 134).

Pots replace these detached fragments of the family. The pots are arrayed, sparkling, on the shelves of the kitchen, like the glittering saris hanging in the metal wardrobe, and for each, there is no limit to what is enough. Above all, by swapping saris for pots, women are maintaining a female exchange economy without referencing men, whatever the financial constraints on the family budget. The *bartan* represent the closure of the deal on a multiplicity of levels.

Used clothing always has a value in one form or another, but it has to be passed on for this value to be realized. This accords with Thompson's model of value creation, which posits that goods must become rubbish in order to be recycled (Thompson 1979). Gifting clothing to inferiors recognizes the transience of leftovers, which are on their way down the scale of value, yet manages to prolong their life by using their social value to make connections of love and obligation and to reinforce hierarchy. Selling clothing as antiques is an attempt to increase its durability as a commodity and obtain a financial return that can be invested in the future. Destroying clothes for their gold content is also a translation of form from cloth to metal that also earns money. Bartering clothing neutralizes its social value in return for symbolically charged pots and allows it to be subsequently transformed in a different value regime, that of the national and international market. Once the initial image is destroyed, the fabric becomes a source for the construction of new images, enabling translation between social worlds. The subsequent use of the bartered garment is the subject of the following chapter.

FIGURE 6. Trading in the early morning, Raghubir Nagar.
PHOTO COURTESY OF TIM MITCHELL

6 ADDING VALUE
Recycling and Transformation

 A piece of fabric holds within its fibers a body's smell, has worn along the creases from folding or sagged in the seat as the wearer has taken the weight off his or her feet; the body leaves its traces on silk, cotton, woolen, and synthetic fibers. Fabrics bear the applied patterns and colors of ethnic regional traditions or are styled into the latest forms dictated by ephemeral fashion. What then is recycled from these elements, and what is forgotten? Most middle-class Indian women believe that the *bartanwale* simply take their unwanted clothing and sell it on to the poor. This imaginary trajectory creates a largely unproblematic fall in value from once-treasured sari, suit, or shirt to anonymous garment, cast off and cast out. Such unwanted clothes appear to have minimal value to all but the most needy, who themselves may be thought to have little value in society. This chapter explores the dynamics of this trade, what actually happens to used clothing once it has been bartered from the domestic sphere and become piles of "stuff" stripped of its unique identities, heaps of material laid out in markets, waiting to be transformed into a new product.[1]

Many clothes do indeed get sold on to the poor and needy, and the poorest Waghri dealers do what they can to add value and make a profit on their steel. As soon as a family can trade up and deal in better-quality used clothing, they are more able to make clever choices in the market, utilizing the elements in the cloth which will appeal to better markets, developing strategies of cultural brokerage to take things out of one social setting and deploy them in another. As Wolf phrased it, the success of the middlemen depends on the separation of buyers from sellers; "they stand guard over the critical junctures or synapses of relationships" (1956, 1075). Middlemen can cause a separation, then bridge the gap and police it;

they have the power to sway popular opinion, condition the way people see things, act as liaisons between social groups and economic communities (Steiner 1995). The greater the middleman's investment and access to resources, the further apart these regimes of value can be and the wider the gap to be bridged, so that recycling results in greater material translations.

The Ghora Mandi

Every morning at dawn, the Waghri dealers sit in long rows stretching across the Ghora Mandi, the wholesale market. The women usually have three or four piles of clothes in front of them, numbering between fifty and one hundred pieces in all (plate 6; figure 6). They almost always sell their own or their families' goods, as they alone know exactly what they paid for them in stainless steel and what they hope to get in return. Most women will have accepted all the clothes offered to them during the bartering process, so long as the final deal was satisfactory,[2] and they try to attract the attention of appropriate buyers as they walk past. Buyers specialize in particular garments, and the women try to tempt them, holding up clothing in the early light. Men's shirts go for Rs 5 to 10, trousers for about Rs 10, and blue jeans for up to Rs 15. Old *dhotis* fetch only Rs 3 to 4. Women's suits are variously priced and rarely have a matching *dupatta;* they are worth about Rs 10 for the worst quality, Rs 15 to 20 for average sets, perhaps Rs 25 for a better one. Suits are generally sold on as clothing, and are worth more if they are not faded, torn, or stained; therefore synthetics are often more prized than old cottons. Party wear is also popular; I saw a brightly colored synthetic *lehenga choli,* made for a special occasion and covered in sequins and rickrack, offered for Rs 40.

Not surprisingly, saris vary most in worth, as they are the most versatile piece of fabric. Old and torn cotton saris can go for only Rs 3 to 10; torn synthetics might fetch up to Rs 15 or 20 if they can be mended, while better-quality synthetics with decorative gold borders might reach Rs 40. Such prices are a complete reversal of the values attributed to them by owners in the upper middle classes, most of whom still shun synthetics in favor of "authentic" handloom cottons. If saris are silk and have decorative borders and ends decorated with *zari,* they are worth much more, even if the field is stained or sagging. The more *zari* on the sari, the more it can be worth, sometimes up to Rs 100 or 150. Some Waghri women claimed to be able to distinguish real gold and silver thread; Ramita said that rubbing real silver on a black stone left a white mark, whilst artificial threads left a red mark, although it was unclear whether this was true. Fine silk saris with significant amounts of gold and silver are now extremely rare in the business; Chanda, a dealer in her mid-forties, claimed to have got one worth Rs 400 ten years before, but had never been offered one like it again.

It was difficult to estimate the average profits of the *pheriwale,* as they range from supplementing subsistence incomes to profitable joint family businesses. The figures here are averages taken from visiting the market several times over the winter months talking to both women and dealers in their homes. Some of the poorest women were handicapped by external factors such as illness and family troubles, reducing both the capital available for acquiring stock and the time they had to go out on *pheriya.* Unimpressive displays of *bartan,* which was all they could afford, meant that they could obtain only poor-quality clothing in return, reducing their profits and completing the vicious circle. A lack of joint family members also reduces the number of people working at home to mend clothing. Other young women claimed to make a daily average profit of Rs 100, a significant proportion of a small household budget. In 2000 Dina, a grandmother in her fifties, had been in the trade in Ahmedabad for thirty years, migrating to Delhi for the winter months with her extended family and renting a small house in Raghubir Nagar. Together with her joint family, she made about Rs 500 a day buying, mending, and selling old clothes.

The label "entrepreneur" is somewhat inappropriate for the poorest Waghri women, for, as Breman has noted, it is a term that hides real disadvantages and unequal power structures in the informal labor sector (Breman 2003). But as soon as women can accumulate a little savings, they can develop their business, and undoubtedly more established or skillful traders can make considerable money. One group I met were setting off to Agra, where the *pheriya* trade was far less developed. A few extended families had got together to hire a big Tata truck, piling the open back high with huge baskets of *bartan* (apparently worth Rs 70,000 in total), children, bedrolls, and stoves. Staying in a temple *dharamsala* (guesthouse), they would scour the better-off districts for clothing, returning several days later laden with garments. On a smaller scale, traders from the areas outside Delhi would also make weekly or even monthly trips to the Ghora

Mandi with their bundles of clothing, hoping to find an urban market for items less desirable in the neighboring rural areas. The *pheriwale's* willingness to travel keeps different styles of clothing in circulation and maximizes the chance of selling them on where they are desired the most, and thus for the best price.

After several months of living in Trans-Yamuna, I got to know the family of *bartanwale* whose patch the Progressive was part of. Once or twice a week, Godavri, a middle-aged mother with a formidable character, would sit on the footpath near the market with one or two of her three daughters, Mira, Bina, and Savita (figure 5). On other days, I would find Mira with her brother-in-law, Raju, sitting on an adjacent road further down, or waiting with a sister-in-law near the spot where the mobile Post Office (housed in a van) stopped every day. No one unconnected with the family ever dealt in their area, and the women never came alone. I met Godavri and her family by chance early one morning at the Ghora Mandi, and was invited back to her home for a drink.

Godavri's house was surprisingly large and comfortable, one of the best in the neighborhood adjoining the slums. It was a two-story concrete building with a room on the roof and a large terrace. Her husband's family had been improving it for thirty-five years, and they had tiled floors and whitewashed walls. Godavri has ten children: five sons and five married daughters, all living nearby and taking part in the business. Their success was evident: as well as continuing to go on *pheriya* every day in the Trans-Yamuna area, her husband and sons had diversified into buying up imported second-hand clothing, which they washed, ironed, and sold on to dealers in Srinigar. Buying blue jeans in Azad Market in Old Delhi, they could guarantee the Srinigar dealers a steady supply, and sold them hundreds of pairs every month or two. Recently they had acquired a bale of "reject" white T-shirts, which they had washed in indigo water in huge vats on the roof to brighten the color. The family was unusual in having entered this trade, which is not traditionally carried out by the Waghri, although more enterprising and successful families are beginning to do so if they can afford to. The women also continued their rounds, thus maximizing their profit in every area. In such a trade, every rupee counts, whether a person is earning fifty paise profit per garment dealing in bulk items or skillfully acquiring a valuable sari for as few *bartan* as possible. By sharing resources and buying equipment jointly, the family was prospering.

From Clothes to Cloth

The buyers in the market are usually men: some whose wives go on *pheriya,* but more often those whose families have struggled up to the next rung of the economic ladder, and who now work at home, adding value to the clothes they buy before selling them on. Their extended families, who often came from the same areas and immigrated to the city together, will often say that they worked the lowest levels of the market when they first arrived in Raghubir Nagar, gradually establishing rights to land and building one-story concrete houses along with their neighbors. Some will have less fortunate relations, others richer contacts, but everyone still living in Raghubir Nagar is working directly in the used-clothing trade.

Such operators work in highly strategic ways, seizing opportunities as they arise and maximizing new niches in the market through personal contacts and sheer hard work. The possibilities for expanding and increasing profits depend upon business skills and acumen and upon a willingness to take risks and forego immediate returns. Another crucial factor is the materiality of the cloth itself, which determines the uses to which it can be put. The relative values of types of clothes are described below, beginning with the cheapest; each material has a fairly established trajectory, but newcomers are constantly attempting to add to the options available and gain an edge over their competitors.

Polishing: Rags and Wipers

The easiest level to begin trading at is that of old cotton saris and *dhotis.* Most *pheriwale* have obtained one or two old pieces in each transaction which cannot be mended and whose value lies in the softness and absorbency of the natural fibers, which makes them suitable for rags, dusters, and polishing cloths. Other old clothes, such as shirts, and bedsheets that are past repair are also turned into rags. In the commercial trade, petty traders need invest little: Gautam bought a few every day for Rs 2 to 3 each, which he tore up into strips and sold as dusters to a local factory for Rs 3 to 4 each, earning a rupee or so on each one. Many of the boys roaming up and down the rows buying *dhotis* and bedsheets work for local wholesalers. Other boys are paid to sit tearing up the clothing: a six-yard sari can become six one-square-yard polishing cloths for the machine industry, *dhotis* are torn into three pieces, and shirts become smaller dusters. Shri Raman Kumar was one such wholesaler, sitting under the covered section of the market and counting up the bundles of cloth amassing at his feet as the morning progressed. He claimed to buy at least thirty large bundles every day, each containing perhaps thirty, forty, or fifty garments, each garment costing from Rs 1 to 10. Once the clothes had been torn up, the bundles could be sold on to hardware shops, factories, and the machining industries.

Some of these men have houses in the lanes around the market, with the ground floor used as storage space. There are over a hundred such dealers around Raghubir Nagar. They often act as middlemen themselves, selling on to others who have large *godowns* (warehouses) where the dealers can accumulate material until they can fulfill larger contracts. Most of the larger dealers are not Waghri and have businesses based in Azad Market (plate 7). Usually Gujaratis or Punjabis, they may be Hindu Vaishyas, Muslims, or Sikhs, and are often refugees from Pakistan. At Independence they were also dealing in old clothing around the railway station, but were quickly integrated into the mainstream economy selling government surplus stock, operating from the municipally established Azad Market.

Trading out of cupboard-sized stalls that are typical of bazaars, these dealers sell on the international rag market. Strips of old clothing are graded by the middlemen into bundles stamped P for the poorest, S for medium quality, and U for top-quality rags, depending on the fabric's strength, size, and fiber content; better cloth is more absorbent and sheds less lint. The biggest players in the market buy up rags from as far afield as Kolkata, Madras, Andhra, and Gujarat, exporting container loads to the Middle East and beyond. They claim Delhi is a major center in the worldwide rag market, turning familiar personal clothing into wipes for paint factories and machine tooling workshops across the globe, all expedited via the *bartan* trade.

Washing and Mending: Recycling Old Clothing

Most of the goods at the Ghora Mandi will become clothing for those unable to afford either new garments in the shops or the second-hand imported clothing available in the local markets. Small dealers specialize in particular garments, depending upon the resources at their disposal to make the items worth more through washing and mending. Raghbir was a typical buyer. Traveling ten kilometers every morning to the market, he bought up to seventy-five old saris every day, which he took home for his wife to wash and darn. He chose mainly synthetic ones, occasionally silk (with no borders), and sold them every Sunday at the Lal Qila (Red Fort), making a Rs 3 to 5 profit on each one. A large part of the market deals in *salwar kamiz;* especially favored are the synthetic suits, which last longer and are easier to maintain. They are often in better condition, not faded but still bright. The range of colors, styles, and fashions of both saris and suits is enormous. The *bartanwale* know what the dealers will be able to resell the garments for, and of course this knowledge significantly affects the early morning bargaining. The dealers will be able to sell on desirable clothing to their customers for a good profit, capitalizing on the fact that end users may buy much better outfits second-hand than they could ever have afforded new. In Indian markets, better-quality used garments from middle-class households provide higher status and better styling, at a lower cost, than poorly made new garments. Most end users will be poor, though dealers often suggested that lower-middle-class women might buy them on the quiet, perhaps telling anyone who asked that they had been gifts from a relative.

A jeans dealer was making a good profit buying a pair for Rs 5 to 10, washing and ironing them, and selling them on for double that price, while another was buying up all the paint-splattered jeans after Holi, the festival of color, in the hopes of making a few paise. One specialized in new and recycled garment labels (figure 2). Another family dealt in old shirts. The husband bought them for Rs 6 or 7 each, and his wife washed and repaired them. He then sold them on for Rs 8 to 10 at Sunday markets across the city. With a turnover of 150 shirts a week, the family could make up to Rs 2,000 a month, a figure that matches the incomes reported by many other traders in the market. Profits per garment are low, less than Rs 5, but the harder wives and daughters in the household work, the higher the turnover of goods can be. These are the families who dry their freshly laundered purchases on lines strung across the alleyways or on the barriers down the center of the divided highways (figure 7). Most of this clothing is then resold in the weekly and Sunday markets found across north India that were described in chapter 2 (plate 11).

Instead of processing high volumes of everyday clothing, some traders specialize in unusual garments for which there is far less competition. Shri Kohli sat at the edge of the Ghora Mandi watching five or six men buying army and police uniforms on his behalf. He estimated that each man usually bought up to one hundred uniforms a week for between Rs 20 and 30. Shri Kohli then sold them on to a dealer in Azad Market, making Rs 5 to 10 profit on each one. He explained that new recruits are issued cotton uniforms for training, but they do not like them and often trade them to the *bartanwale,* buying themselves cotton-synthetic ones that are easier to maintain. The tatty, mismatched uniforms are subsequently sold in Azad Market and are bought by *chowkidars,* private guards such as those at the Progressive, who have to provide their own clothing.

Every potential market is explored and exploited. Shri Gupta is a Bihari trader who began making skirts out of worn *kurtas* for the Assamese market several years ago. Women and children working in the tea plantations there habitually wear ankle-length skirts and *cholis,* not Punjabi suits or saris. He buys between sixty and seventy suits a day, each consisting of a *kurta* and a pair of *salwar,* at about Rs 10 for each suit, and employs two tailors to stitch his designs. He cuts the top of the *kurta* off just below the armholes, across the chest, leaving the skirt part intact. As this is usually too short to make a long skirt, he then undoes the sleeve seams and reuses the sleeve fabric for a waistband by turning it horizontally and stitching it across the top of the skirt, one sleeve at the front, the other at the back: the lengthwise grain of the fabric runs properly across the waistband, strengthening the garment. Finally he folds the new top edge over, forming a tube for the waist elastic. The skirts are sold to an Assamese dealer for Rs 12, the *salwar* fetch Rs 3 to 5, and the leftover bodice and back pieces can be sold as scraps, *katran,* to the local rag merchants.

Moving Up: Trading Silk Saris

At 3:00 AM every morning, the most valuable saris are changing hands in the Ghora Mandi. Women place silk saris and those with decorative *zari* borders on the top of their heaps, shining in the half-light, and the earliest buyers snap them up. Prices vary according to the size of the decorated surface and the amount of gold and silver used in the borders and *pallu.* One or two buyers call out for those containing real gold and silver, paying a few hundred rupees for them and later taking them down to the Kinari Bazaar to be sold or burnt for their metal content. Very occasionally clothing of extremely high quality is offered, and some people suggested that it would be resold as new through select outlets, but I found no evidence of this. Unwanted silk saris are usually used as raw material to be refashioned into soft furnishings and tailored clothing.

All of the buyers are middlemen, building up stocks in their homes around the periphery of the market to fulfill contracts with large manufacturers or selling on directly to the family businesses in the neighboring Paschim Puri district that transform them (plate 12). Most buyers acquire as many as they can afford to invest in every day; although each *pheriwali* claims to receive the finest silk saris only rarely, common estimates suggest that at least ten thousand people go out on *pheriya* every day, so there are always plenty available. These garments are the kind the women in the Progressive wear to dress up, and they are often part of a trousseau or received as gifts later on. When they appear in the market they are usually high quality, but irreparably stained. When originally bought they can cost from a few hundred to several thousand rupees; one buyer suggested that a Rs 2,000 sari probably sells in the Ghora Mandi for about Rs 100, 5 percent of its original price. The *bartanwale* will be able to make several rupees' profit, depending on their bartering skills.

The buyers in the Ghora Mandi tend to earn about Rs 15 to 25 per silk sari if they sell them on in bulk, although some highly decorated individual pieces would earn more. There are only twenty to twenty-five silk sari dealers in Raghubir Nagar, who try to compete with the sixty to seventy more established dealers in Paschim Puri. One pair of brothers, who sold to a manufacturer in the Okhla Industrial Zone, claimed to make a monthly profit of Rs 20,000 to 25,000, supplying bundles of five hundred saris at a time. Another dealer, starting out alone, was just managing to make Rs 8,000, and his wife continued on her *pheriya*. He found it hard to compete with the buying power of the larger dealers, and was often forced to sell his saris to them for a lower profit.

Few lower-level dealers specialize in types within the general category of "silk saris," and houses bulge with multi-colored bundles of saris in every style: satins, tissues, tie-dyed *bandhini,* Banarasi brocades and lighter *tanchois* brocades, and south Indian silks. One or two have found niche markets for particular types. A Muslim trader from Aligarh, now living in Raghubir Nagar with his three brothers, specialized in printed silk saris which he sold to a manufacturer in Naraina. His father had been in the trade for forty years before him, and he himself for eighteen years. Those with the tiniest floral sprig were worth the most, up to Rs 100 each, whereas larger patterns fetched only Rs 50 to 60, being more difficult for pattern cutters to make new products from. He bought three to four hundred a month, and other dealers came to him when they acquired them.

He had another sideline in "airport saris" (worn by Indian flight attendants), which he was particularly keen to locate, although he was reluctant to explain their particular attraction.

The most successful dealers in Raghubir Nagar are those who have acquired influence and control in a variety of interconnected fields, eliminating the middlemen. Sushil was in his early twenties and beginning to deal in silk saris. Keen to meet tourists with whom he could deal directly, but who rarely found their way to northwest Delhi, he had filled his house with the most gorgeous silk saris, many of which he offered for several hundred rupees each. He claimed to use family contacts to bypass the local Ghora Mandi and obtain clothing directly from dealers across Rajasthan, extolling for my benefit the quality and beauty of saris formerly belonging to the royal princesses of minor desert courts. Undoubtedly he was able to exploit such extensive contacts, and did not rely solely on sending scouts to the market every morning. His father was the president of the Bartan Traders' Association and a prominent member of both the local *panchayat* and the administration of the Ghora Mandir temple, so the family had influence and contacts in every sphere of business undertaken in the district. Sushil was making a name for himself as a dealer, and women sold to him directly. However, the market system allows the poorest sellers to play buyers off against each other and new players to enter the field.

Furnishings and Fashions: Targeting Western Markets

Bedspreads and Cushions

A short cycle-rickshaw ride away from Raghubir Nagar, the suburb of Paschim Puri is a definite rung up the ladder of success. Houses are three-story brick buildings with roof terraces, courtyards, and marble-chip flooring. Streets are arranged around leafy squares, and cars and motorbikes line the pavements. The area is modestly prosperous, comfortable, middle-class in character. Using introductions from some of the silk traders in the Ghora Mandi, I tried to establish what actually happens to the thousands of silk saris siphoned off from the market. It soon became clear that these dealers were operating at a multitude of levels simultaneously, as buyers, manufacturers, sellers, and exporters of soft furnishings made entirely from recycled clothing.

The Paschim Puri traders are all Gujaratis, who have been developing the business of recycling Gujarati embroidery since Independence.
Tarlo documents the manner in which Gujarati embroidery was traded
for *bartan* across the states of Gujarat and Rajasthan and sold to collectors
and foreign tourists (Tarlo 1996b). She follows the uptake of such peasant
craftsmanship by the urban elite in Delhi, who made "ethnic chic" *the* look
of the 1980s and 1990s (Tarlo 1996a); it then trickled down in the form
of acceptable fashion for the Indian middle classes. Investigating the Law
Gardens market in Ahmedabad, Tarlo showed how the Waghri dealers are
selling contemporary products made from embroidery to the local middle
classes and foreign tourists, and are now producing machine-made embroidery to cope with the demand (Tarlo 1997).

The Delhi community of the Waghris is part of this network of dealers and manufacturers that extends across west and north India, and most
of the people I spoke to were still using new and old Gujarati embroidery
to make new products for the export market. The whole extended family is usually involved in manufacturing, and many employ several tailors
on a piece-work basis. The most common products are cushion covers,
bedspreads, and wall hangings made from scraps of embroidered clothing. Coarse black backing fabric is cut to the desired size and pieces of
embroidery are laid out over it in a design. Although square patches are
occasionally used, the current fashion is for irregular pieces: the overall
effect resembles crazy paving. PVA glue is then pasted across the surface
and the pieces attached. Edges are not turned over, and little sewing is
needed. Instead, long hanks of twisted cotton strands are laid down over
the joins and couched in place. Whilst some tailors try to create appealing color combinations within overall designs, a more recent trend is to
mix together whatever is handy and overdye the whole piece, creating a
color wash effect in yellow, red, green, blue, or grey. This enables scraps
and pieces to be used even if their colors clash, a thrifty practice which has
proved popular with foreign customers.

In the last fifteen to twenty years, these manufacturers have turned
from primarily using Gujarati embroidery to using the growing abundance of decorated saris to be found in the nearby Ghora Mandi, creating
an extended repertoire of designs for soft furnishings, each determined
by the style of the sari (plate 8). Silk saris are cut up into pieces, according
to their basic construction: the top and bottom borders (about five yards
long), the decorated end or *pallu,* and the central field, often containing

regularly repeated motifs. The cloth's construction, weave, and texture are considered, along with whether borders are striped or tapestry-woven with motifs and whether design elements in the *pallu* and field are small and repetitive or big and bold. Although the saris were initially selected for their material properties of silkiness and *zari* content, it is the particular combination of design elements used in each type of sari which determines their reuse.

Banarasi brocades are the most commonly used, as they tend to have bold *zari* borders with definite stripes, heavily patterned *pallu* in *zari,* and regularly recurring flower *buttis* (decorations), animals, and geometric shapes in the field. The remaining, plainer silk parts of the sari may be sold to others to be made up into scarves (*chuni* and headsquares), used as lining material in recycled clothing, or, if in poor condition, sold for scrap (*chindi, katran*) to the local rag merchants. Cushion covers and bedspreads are made up of a patchwork of these elements stitched together on cotton backing. A popular design is of a series of concentric bands of sari borders around a central square, laid down diagonally within the square or rectangle of the object. Borders of different colors and *zari* patterns are mixed together to create dazzling effects. Similarly, two or three saris with contrasting colors and motifs may be cut into squares and arranged into a well-designed overall pattern. In plate 8, a transparent white cotton *jamdani* sari with elephant and peacock motifs in black has been paired with a red and green silk sari with heavy silver *zari* borders. This double bedspread took six days to make, and used pieces from twenty sari borders and one central field to make the three hundred and four squares contained within.

The Paschim Puri traders are now some of the main manufacturers of these products in north India. The Ghora Mandi cannot supply enough silk saris to meet demand, and so the traders have capitalized on extensive networks across north India to buy up more raw materials and sell on their finished products. Ashok Kumar was one such trader, and he has buyers working for him in second-hand markets in Ahmedabad, Mumbai, Surat, and Banaras. The latter are both centers of production and good sources for *zari* borders. Smaller traders in small- and medium-sized towns across India sell on their used clothing to middlemen, who lug heavy bundles of second-hand silk saris across the Indian rail network to reach the major cities and finally Delhi. There they are made up and sold back to traders in tourist destinations such as Goa, Mumbai, Jaipur, and Agra or exported to Nepal and beyond. In Delhi as elsewhere, petty traders, usually Waghris, sell a few cushion covers outside the main markets and peddle them on the main streets, such as Janpath and Connaught Place, while shopkeepers make a living in the poorer parts of town, such as Paharganj, selling them as souvenirs to backpackers (plate 13). One dealer suggested that a sari cushion cover might wholesale for Rs 20 to 25, but tourists on Janpath might pay Rs 50 to 60; however, new arrivals unwilling to bargain probably part with many times that amount.

These products are ubiquitous, on offer to foreign tourists wherever they are to be found across India and south Asia as a whole, and they are now being imported in vast quantities into the West by entrepreneurial Western travelers-turned-traders. The spread of such patterns across the surfaces of Western households in the form of hangings, throws, and cushions creates a fabric of Indianness in daily life, and presents images of incorporated, mutated alterity to the inhabitant.[3] Sold through "ethnic" trading shops, market stalls, and festivals, these contemporary recycled items have become hugely popular. They are associated with a free-floating aesthetic repertoire with an Indian or, even less specifically, Asian style, a mixture of colors, silks, and patterns that look Indian but that the customer may or may not associate with saris. However, until recently, they were not marketed as recycled. This has changed between the time I started my research, in 1999–2000, and my writing of this book in 2009; the entrepreneurial necessity of manufacturing goods from cheap recycled materials is now promoted to customers as recycling, though previously it was unmentioned.

Like the producers of many products developed to recycle cheaply available resources, the suppliers of recycled sari cushions rely on what is available *ad hoc,* and the saris, brought in by the Waghri traders, constantly exercise their creativity. The overall style of "Indian" ethnic sari products is now well defined, and their producers understand the potential of design elements in various types of saris and exploit them through forms such as patchwork and stripes. However, each of these pieces is a unique combination of original materials that cannot be exactly replicated.

Recently, Indian silks and textiles have emerged as a major fashion trend in Western mainstream design, of interior furnishings as well as clothing. This has led low- to middle-market UK chain stores to sell "sari cushions," featuring identical diagonally striped designs, modeled after patterns originally developed to reuse sari borders. The 2001 catalogue *The Pier* described Pier One's "sari cushions" as "made from Indian sari borders . . . made from individual pieces of fabric" and warned, "Patterns may vary." Each was priced at £29.95. However, large retailers in the West such as Pier One cannot rely solely on serendipity and happenstance: they need to be able to guarantee consistent quality and large quantities of similar, if not identical, colors and designs to support their seasonal marketing strategies, catalogues, and displays. New sari borders are bought in bulk and made up to copy the "original" recycled product, guaranteeing thousands of cushion covers that have no tears or pulled threads, and no chance of a surreptitiously hidden stain in a corner.

The reuse of clothing as furnishing fabric extends habitual domestic practices into the commercial sphere, and it is no surprise that its origins as a recycling practice are lost once international profits are at stake. Sari borders have always been saved in the practice of thrift, and either sewn onto new plain silk saris or used to decorate suits. The aesthetics of patchwork as a practice and the kind of objects made with it, such as cushions, throws, and covers, have long been common to Indian and Western culture, although commercial products are undoubtedly geared toward export. However, remaking silk saris into Western fashion garments involves a conceptually more radical transformation. To investigate this transformation, I take the example of the markets in Pushkar, a small town in Rajasthan, where this practice, a modern variation on an old theme, is said to have been most recently revived.

Travelers' Clothes in Pushkar

The holy town of Pushkar, with the crumbling lakeside palaces that once belonged to the maharaja and its plethora of temples, has long been regarded as a haven by Western backpackers, a sanctuary from the rigors of budget travel across north India. Many spend days or weeks meeting up with friends, relaxing, and whiling away the hours eating comfort food and shopping for cheap souvenirs. They are lured in part by the beguiling descriptions found in independent travel guides of an ancient Indian sacred place coexisting with a hedonist consumer paradise:

> With its smooth spread of white-domed houses and temples reflected in a tranquil lake, Pushkar . . . resembles nothing so much as a

pearl dropped in the desert. . . . No one knows quite how old Pushkar is: legend relates that at the beginning of time, Brahma dropped a lotus flower (*pushkara*) from the sky, declared the lake that sprang from the arid desert sands to be holy, and promised that anyone who bathed in it would be freed from their sins. As the site of one of only two temples in the world dedicated solely to the Lord Brahma, Pushkar attracts a constant flow of pilgrims to worship and bathe from its fifty-two lakeside *ghats* . . .

The streets of Pushkar are too narrow for traffic, and there are no rickshaws; a lazy stroll around the shores of the lake and through the main bazaar takes little more than an hour. Shopping is the prime pastime for visitors; you're encouraged to take all the time you want browsing at the countless small roadside stalls, choosing from a vast assortment of clothes, jewelry, second-hand books, *chillums,* Indian classical music CDs and cassettes, and Rajasthani paintings. You'd do better, however, to buy the latter in Udaipur; clothes here (especially of the hippie variety) are usually much better buys. (Abram et al. 1996, 165–66)

Clothing is displayed in abundance, arrayed across every shop front and adorned with fluorescent paper labels proclaiming its bargain prices. Every conceivable style of "ethnic," "Indian," and "hippie" clothing is on offer, often unisex, cheaply made out of Indian cottons and synthetics. There are loose drawstring trousers, shirts, smocks, tunics, skirts, hats, and bags, made from Indian-style block prints and naturally dyed cloth, tie-dyes, plain creams and whites, and bright shades with star, moon, and sun motifs. Winter jackets in felt and wool with Nehru collars and Nepali braid trimmings hang alongside piles of Rajasthani woolen shawls and synthetic sweaters. Every taste is catered to; skin-tight Lycra tops and hipster bell-bottom pants are also for sale, often displaying hip Western logos. Much of the clothing is simple to make and one-size-fits-all, thus easy to sell.

Many of these clothes are bought to be worn immediately and thrown away as they wear out on long trips across the continent. The buyers may not favor "hippie" style when they are at home, and donning them can be an important part of the travel experience. Other clothes are definitely in the latest fashions and proclaim their foreign identity. The highly disproportionate number of backpackers in the town has resulted in an overlapping commercial and social sartorial arena along the main street in Pushkar, where it has become acceptable for Western tourists to wear revealing clothing. The mass influx of such tourists to Pushkar since the 1960s has led to the growth of a niche market for "cross-over" products, which feature Indian materials but are developed or adapted specifically to cater to the limited budgets and particular tastes of the visitors.

Many of these visitors never venture beyond the few hundred meters of the main street. But Pushkar has become a major manufacturing site behind the scenes, with entrepreneurial Indians recognizing fashion trends brought in by the visitors and making products locally that can easily be updated. One of the major ways in which new styles are introduced is through direct copying, often initiated by the travelers themselves. Amongst them, tailors in particular have a good reputation for their ability to pattern new clothes on favorite garments in Western styles, using local materials such as silk and cotton. Indeed, prospective visitors to India are often encouraged by friends who have been there to take favorite clothes with them specifically for copying by tailors. These hybrid clothes re-create travelers' wardrobes at a fraction of their cost in the West, with the added bonuses of being both one-of-a-kind garments and unique combinations of Western style and Indian fabrics. For a generation of young Western travelers who are largely unused to made-to-measure clothing and who normally buy everything ready-made in chain stores, it is immensely appealing to have garments made especially for them, so that they can alter the length and fit as they like and choose the pieces of cloth to be used.

One of the most successful types of clothing on offer are Western-styled garments created from saris. The elements of a sari have always been available as a resource for Indian women creating home furnishings and *kurtas*. Now the same elements—borders, ends, and fields—are being used for shirts, sundresses, skirts, shorts, halter-neck tops, pedal-pushers, and drawstring trousers. Sari borders are used as necklines, sleeve edging, and hems; the *pallus* serve as panels, bodices, trouser legs, and waistcoats; and the field makes up the bulk of the garment.

Two or three retailers claimed that such clothes were first created in Pushkar, fifteen to twenty years before my research there. An "origin myth" says that a young Western woman took her dress to a Pushkar tailor, together with an old sari she had got hold of, and asked him to stitch a copy for her. The tailor picked up the idea and ran up a whole batch that sold immediately; he then opened up a shop. Some tailors were used to making up *kurtas* out of saris and understood how the formal properties of the sari's components could be adapted and reused. With this knowledge added to their well-established copying skills, it was relatively easy to make the new garments. The number of shops selling them proliferated, and manufacturing units in nearby houses were set up. One shopkeeper estimated that there were at least sixty such units in Pushkar and Ajmer making these clothes for sale across India and for export to the West. Certainly Pushkar is the center of sari clothing manufacture across north India. A ready clientele, bringing in new styles from abroad, keeps the transformed garments up to date with changing fashions.

Although Pushkar could provide the labor to make these clothes, there was no ready supply of saris at affordable prices. The Rajasthani peoples in the outlying area wear full skirts, blouses, and large veils. The town itself is small, and it and nearby Ajmer together could not provide enough second-hand clothing. However, links between locals and networks of Waghri traders have ensured that Pushkar is now a major market for many old silk sari traders across north India, with agents visiting the workshops to pick up orders by the thousand. In Delhi, some of the Paschim Puri traders sold to manufacturers in Pushkar, and one smaller trader in Raghubir Nagar took the train there weekly with his single bundle, hoping to get a better price by evading the competing Paschim Puri dealers. When I visited Ahmedabad just before the holiest festival in Pushkar, Karttika Purnima, I met several women who were stockpiling thousands of saris, planning to take them by bus to the dealers in Pushkar and make a pilgrimage at the same time. One woman had been buying up silk saris in Mumbai over the previous months in preparation for the festival, while others used networks of local petty *pheriwale* to amass stock.

Most workshops are not interested in buying the highest-quality silks, and may pay as little as Rs 35 to 40 each if buying by the thousand. Stains, holes, and tears are unimportant, as local pattern cutters are adept at avoiding them. In fact, these saris are never washed or dry cleaned throughout the whole transformation. Whether a sari is used in large pieces or becomes many smaller decorative parts of larger garments depends on its condition, design, and quality.

I spoke to Ashok Bhai, of the Gurudit Garment Shop, who had twenty tailors working for him; his whole family was in the business, which he started about fifteen years ago, and he had five shops, although the other four sold new Indian clothing (saris and *salwar kamiz*). He claimed that his Western customers did not seem to mind if the clothing was torn or dirty, unlike his Indian clientele, who want only "good, clean, new things" (implying that they would not touch potentially polluted cast-offs). As we were chatting, two Indian girls in jeans, T-shirts, and sunglasses stopped for a moment to look along his rail of skimpy sari clothing on the street, apparently also unconcerned about its condition. Ashok was sure that these "Mumbai City girls" wanted the Western look on the cheap: "they want to show off their body," but would probably not tell their parents where the clothing came from or what it was made from. Their curiosity satisfied, they wandered on without buying.

Cathy, a British woman in her early thirties, had bought various garments during her travels in north India. She had acquired a couple of plain cotton shirts to wear with jeans from an export surplus stall in Delhi, and an embroidered top that was an export reject from the same market. Staying with a local friend while in the capital, she had had a very plain maroon *salwar kamiz* made up, noticeably more subdued in design than her host's was likely to have been, with no trimmings. The gold *zari* borders and neck embellishments commonly found on such clothing were "too much" and "too Indian," she said, and on an already unfamiliar garment she thought they would have made her feel self-conscious. Now she had met up with friends in Pushkar and was staying by the lake for a week to "get away from the hassle of Delhi." She had been tempted into buying some of the 1970s retro hippie clothing on offer, including a long orange dress which she'd worn once while there. The vibrant color was something she felt she would never have worn at home, though she may have wanted to, and she connected its bold hue with being in India, and specifically with being in Pushkar, a tourist venue whose freedom she was enjoying. When we met she was in a small shop buying an emerald green wrap-around sari skirt with pink and gold *buttis* and a wide gold thread border. She liked the fact that it was an easy style to wear, was lightweight, and could be shoved in her backpack; she also thought that the colors were great, and not ones which she would easily find at home.

In fact, because many seasonal fashion ranges in the West offer only a limited palette of colors and designs, the vibrant colors and patterns of saris were particularly appreciated. Cathy didn't mind a slightly grubby patch on her skirt, as she was only paying "a couple of quid" for it, she thought the stain would wash out, and she didn't think she would ever wear it at home anyway.[4] Later she and one of her friends talked about the colorful clothing Indian women wear, and how the sari clothes gave them the opportunity to wear some of these fantastic fabrics and bright shades without feeling as though they were imitating Indian dress. Yet they did not expect to wear them again at home, where they might seem odd or too "ethnic." The pieces of old silk sari enable women to experiment with images and color, to play with the boundaries between the exotic and the familiar, in a parallel social space that is constructed largely by and for the enjoyment of travelers in the heart of the Rajasthani desert.

Clothes made of old saris are cheap for tourists; retail prices in Pushkar are relatively low, and trousers go for Rs 50, sundresses for Rs 70, and long dresses for up to Rs 100. Clothes made with fabric taken from new saris would be far beyond the price range of these travelers. Bolts of brocade cloth are also more expensive, and they lack the *pallu* and borders of the complete woven sari and therefore the design possibilities afforded by mixing and matching pieces. Using saris also ensures that each garment is unique: not even skirts made from the same sari will ever look the same.

The Western fashion garment created from used silk saris raises complex questions concerning authenticity and appropriation, identity and individuality. The sari itself was a unique piece of handwoven, decorated fabric; each sari is an original, conforming to regional styles or fashionable designs, both its form and its style identifying its Indian origins. These recycled clothes are incorporated into the wearer's identity; many travelers, who are undergoing many changes and new experiences, acquire and dispose of clothing rapidly, often wearing out clothes as they leave the culture that produced them and buying new ones as they arrive in a new place. They may discard clothes easily when moving on, because of overfull backpacks, changes in climate, or sheer boredom.

These customers, both men and women, are dressing up, re-creating themselves performatively as they travel as backpackers on the road, far away from home, and this re-creation is mirrored to some extent in their clothing.[5] They are enjoying the experiences offered by the society through which they travel, but are to some extent isolated from the place in which they find themselves. There is a correlation between the hybrid sari clothing they purchase and its wearers' own situation. These travelers are wrapping themselves in different guises, in fabrics, colors, and designs that proclaim their marginal identity (not Indian, yet not quite typically Western either), but are in some sense also claiming to belong everywhere.

Like all clothing, the hybrid sari clothing also transforms the body of its wearer through its unusual cloth or cut. It is often looser, baggier, and freer than is usual in recent Western fashion, reflecting a more laid-back, relaxed lifestyle; yet it may also be tighter and more revealing than conservative north Indian sensibilities would normally approve of, thus both creating and displaying the tensions of many travelers' experiences, especially women's. It is striking that amidst the apparent diversity of choice, there are in fact only minor variations on a few major themes in what is on offer in the hundreds of small shops and stalls. Because of this, wearing such clothing means declaring oneself part of a group of Western travelers, and this is as important as any unique qualities of a garment commissioned.

These clothes fall into the category of tourist-objects defined by Lury. Lury identifies three types of objects of travel. Traveler-objects retain their meaning across contexts, and retain an authenticated relation to an original dwelling, while tripper-objects have meanings that appear arbitrary, imposed by the outside through external context or final dwelling place. Tourist-objects lie "in between there and here in their journeying"; they are neither closed objects whose integrity relies upon their relation to an original place, nor open objects whose integrity is eroded by their final resting place (Lury 1997, 78–79).

The ephemerality of such clothing and its susceptibility to singularization through wear and tear constitute part of the travelers' experience of journeying. These qualities set such tourist clothing apart from other trans-national commodities discussed under the rubric of "tourist art."[6] It is this potential transience which cross-cuts its authenticity:

> Just as authenticity in an artefact represents stability and longevity by referring to the perceived moment of origin some time in the far distant past, its antithesis, ephemerality, places emphasis on the present, that fragile evanescent moment that passes in a flash. (Attfield 2000, 81)

The materiality of the cloth itself is ephemeral. It has been cast out once and rescued; already tatty, it can now be picked up and worn for a while.

For these clothes are markers and makers of journeys; they take on the shape of the new owner's travels.

> Ephemerality . . . offers a condition resistant to closure and materialises uncertainty. . . . Authenticity and ephemerality can be said to materialise the relation between time and change in ordinary things . . . [and it is their] particularly adaptable character that makes textiles analogous to the provisional nature of the contemporary sense of self-identity. (Attfield 2000, 87, 88, 132)

The hot pants, halter-neck tops, and Capri pants tailored from old wedding saris and family gifts hold the tensions of continuity and change within the very structure of the fabric. Simultaneously at risk of wearing out and falling apart and of being deliberately thrown out again, their hold on life is tenuous.

These decorative silk fabrics have the potential to translate images of the traveler's transience and impermanence through their own adaptability and change in form, while enabling various nuanced perceptions of belonging. Feelings of association simultaneously allow the clothing's wearers to express their individuality, to fit in with other tourists through the creation of a specific sartorial culture, and to reference at a distance the host culture through which they are traveling by the reuse of local aesthetics.

Delhi Workshops: Mrs. Sharma

Not surprisingly, the popularity of such clothes amongst travelers and their relatively low production costs have spawned a flourishing informal export trade. Entrepreneurial travelers order stocks from the workshops in Pushkar (and in other centers, such as Banaras, Delhi, and Mumbai) at very low wholesale prices to sell back home. Ashok Bhai, no doubt still hoping for an order from me, quoted in the range of Rs 30 for a sundress, Rs 50 for a long dress, and so on, and displayed an order book listing clients from across the world as evidence of his ability to supply quality clothing. The handwritten orders were often completed by the travelers themselves: "John and Vicky from Bath, England," "Peter, . . . Germany," and others from Japan, Singapore, Israel, the U.S., and Australia. Because it was made from used material, such clothing did not require the Indian manufacturers to obtain export licenses, which were not easily available to small-scale businesses before the recent relaxation of the law. Ashok Bhai argued that the clothes were particularly attractive to English traders because, he claimed, "second-hand clothing" is not subject to a high customs tariff when imported.

Once imported, these clothes are then sold on in the informal sector: at music festivals, in street markets, and in independent "ethnic" shops across the West, often in conjunction with cushions, bedspreads, jewelry, and other decorative items. In England in the early 2000s they generally sold for between £10 and £30, depending on their quality, the retail site, and what the market would bear. But as Ashok Bhai pointed out, although he is careful to chose unstained saris, his buyers cannot chose color schemes or design features—these depend solely upon the saris purchased, and therefore each garment is designed by the Indian tailors in the workshop. The garments are therefore reclassified by their potential consumers: no longer personal souvenirs or travelers' gear, they take on the nature of imported exotica, undoubtedly invoking different perceptions of authenticity and identity.

Back in Delhi, the international trade in recycled Indian clothing was moving upmarket. My fieldwork coincided with the peak of the West's taste for Indian-influenced clothing. By mid-1999, British stores were selling mass-produced clothing featuring embroidery in a "Gujarati style," appliqué designs using Indian fabrics, and jeans with *zari* borders attached, as the popularity of the clothing found in street markets across the country trickled up the status hierarchy. This taste was no doubt also fueled by the growing number of backpackers and clubbers traveling to India directly who were buying such clothes in travelers' havens such as the Anjuna flea market, the Kulu valley in Himachal Pradesh, and Pushkar.

Some forward-thinking Western entrepreneurs realized that they needed to ensure that the niche product they were selling was of reliable quality in order for it not to be swamped by the cheap clothing available in the global chain stores, which would wipe out their profit margins. In addition, once the overwhelming fashion craze passed, there was a danger that tastes would change and the recycled sari dress appeared "tired."

The traders in Raghubir Nagar and Paschim Puri who deal in old silk saris sell in bulk to manufacturers around the city, but are extremely reluctant to reveal details. However, eventually I was directed to Mrs. Sharma of Pummy Traders, who maintained an office and small workshop on the second floor of a crumbling office building near Connaught Place. Mrs. Sharma is an extremely shrewd Gujarati businesswoman in her early sixties; she also likes to style herself an entrepreneur. She had, she said, not a single Indian customer in the whole used sari business, and dealt only with foreigners (thus promoting herself as a cultural broker *par excellence*). Among her partners is the British company Tommy Frog, which works with her to produce a higher-quality range of clothing.

Her own "origin myth" began with her moving to Delhi, newly married and impoverished, and living in rented rooms in Paharganj, by the New Delhi railway station. Although she had four children (and now ten grandchildren), she had an adventurous spirit, and claims to have set off for the U.S. in 1976, sending ahead three hundred kilos of jewelry, raw silk, material, semi-precious stones, and handicrafts. Because she was a middle-aged woman, no one had wanted to invest in her, and she had to work hard to raise the funding for her first trip. Buying a 1966 Chevy, she traveled across thirty-four states buying and selling jewelry and crafts, sleeping in her car for six months, before returning with $20,000. In 1979 she returned for a longer period, opening a small shop for a while, before establishing herself permanently in Delhi as a manufacturer and exporter of clothing (plates 14 and 15). In the early 2000s her son took over much of the daily running of the business; he himself had traveled to Europe twice in the late 1990s to deliver goods to customers and get more orders, although he spoke no European language except English (with a heavy Indian accent) and had had little money. He recalled once sleeping on a train platform in Italy, his head pillowed on his bag of sari clothes, while waiting for a client who never materialized.

Mrs. Sharma tells her story to encourage the traveling backpackers who tentatively find their way to her office to identify with her, to believe that they too can succeed through selling Indian products abroad, and to fill them with confidence in her ability to help them. Because she remained unconvinced by my efforts to deny any interest in joining her in a business venture, she went on, claiming that she worked only from the heart and that the clothes she made were "divinely inspired, as is the work of all true artists." Westerners' recognition of this is, she said, why Indian clothing had seen a renaissance in the last few years. "Gujaratis have a particular art—they are not educated, and their imagination works through cloth, not paper," she explained. She had been in the business of making these clothes from old saris for the last twenty-five years and had seen fashions such as punk and "gay" come and go—but "life itself is a recycling, and people make the world beautiful themselves." She said, "I see things, copy them, change them slightly, and make something new."

She said all this over a desk littered with faxes of detailed designs and patterns of orders sent from London. In fact, she works closely with her buyers, who return home to Europe, the U.S., or Australia and send her repeat orders, with new fittings, alterations, and suggestions, as they develop new ideas of what will sell there. One of the biggest difficulties for such customers is quality control, and at least two that I met had had to return to Delhi in frustration to deal with poor-quality, incomplete, or terminally delayed shipments. Mrs. Sharma has a network of agents who buy silk saris for her in Raghubir Nagar, Mumbai, and Pune; she proudly declares she can get hold of five thousand in a day if necessary, but the manufacturing side of her business leaves some customers in despair. Many of the clothes she makes are stained or badly sewn, and can be worth only a few pounds in a British market. Still, she has clients from all over the world who return to her, and the profits and lifestyle they enjoy are obviously good enough for them to continue doing business with her for a while.

London's Camden Market: Tommy Frog

Mrs. Sharma's buyers face increasing competition in the West as a result of the popularity of the type of clothes she and similar manufacturers produce. Yet she knew that the time would come soon enough when they would no longer be in fashion. She therefore went into business with an English couple who were determined to take the products upmarket and invested in her manufacturing unit. Simon and Sarah Wilson spent nine months traveling around India and Nepal on a Vespa, returning to the UK with a bagful of jewelry that they started selling at weekend markets. As Indian clothing started "taking off," Simon returned to Delhi to invest in some new stock, meeting Mrs. Sharma. A former TV cameraman, he was an adept businessman who clearly enjoyed dealing with Indian traders; his wife, Sarah, designed the clothes he commissioned, although she had had no formal training. According to Simon, "Mrs. Sharma is Shakti [the great goddess, a personification of female power]—she bullshits, but she gets away with it."

After initial success in Camden Market on weekends, the Wilsons both decided to give up their jobs and invest heavily in the business, which they named Tommy Frog, leasing a shop in Camden Lock in order to avoid "getting stuck" in the street market. Returning to London a few weeks after meeting Simon in Delhi, I visited them there. They had smartened up the shop themselves, changing what had been a typically dark, "hippie" Camden shop to a light and airy space painted white and fitted out with IKEA shelving, glass-topped coffee tables, and a comfy sofa. The windows were dressed with shimmering gauze saris in a range of pastels and displayed a dazzling array of evening bags in rainbow colors. The effect was modern and smart, yet also managed to be glamorous and sensual, attracting better-off customers in search of something new. The clothing was hung and draped artfully, rather than squashed onto the rails, and customers were encouraged to browse and chat.

However, the most noticeable change from their presentation in the market was in the clothing itself. Instead of relying solely on the saris that had been available to Mrs. Sharma in the Indian market and her tailors' skills, they had created a unified effect for their range by choosing a basic palette of plain pastel silks and using them as linings and trimmings. The same plain fabrics were made up into matching plain skirts and tops, moving their stock away from the street-market Indian/ethnic/hippie look to a more understated exoticism. Although they still use some synthetic saris with artificial pearls and poor-quality *zardozi,* the clothes are stitched more carefully and none have stains or tears. As a result, the clothing retails at much higher prices and is no longer associated with backpackers, having a little more sophistication.

In fact, Simon had his own version of how Indian clothing and materials had succeeded in the UK market, in which Tommy Frog had played a part. He recounted how Indian recycled clothing had been extremely popular at the festivals during the summer of 1998 and had quickly moved into street markets, where he and his wife had identified it as the next big fashion. The clothing had already attracted the attention of the glossy fashion magazines and upmarket fashion retailers; he attributed this to Maggie May, a dealer in Portobello Market (involved in the vintage designer clothing shop Rellik). At that time she had started sewing sequins on the edges of her jeans. Bianca Jagger had begun wearing them, and the jeans became high-fashion items, retailing at £150 to £200 each. By the summer of 1999, elements of "Indian" fashion were to be found in much street clothing, and larger fashion labels such as Miss Selfridge, which regularly send their scouts out to Camden and Portobello Markets, were catching on. At the top end of the market, Dolce & Gabbana brought out an embroidered silk belt retailing for several hundred pounds, which Simon told me was featured in *Tatler* as a "glorious little belt." However, he claims it was all too similar to the cummerbund-style belt his wife Sarah had designed and put on sale in Tommy Frog; although perhaps a little galling, the copying could only be good for business.

The fabrics of contemporary recycled textiles may have qualities that have a long tradition in India. Yet they often contain colors, designs, and patterns which have moved in and out of fashion within India, sometimes in relation to recent international trends. Western styles made from recycled saris are hybrid in a multitude of ways, from fabric design to form, but as cloth and clothing they have an intimate, ephemeral nature that allows them to be singularized in a particularly personal way. The wearer is brought into the global arena through the connotations of glamour and exoticism woven into the fabric, in a very different manner from a person simply wearing a Western-style garment manufactured in a sweatshop in the developing world. The silk sari is both destroyed and transformed into a halter-neck top and pair of hot pants, radically highlighting the different systems of moral values mediated by the cloth itself and its relationship to the female form in particular.

The recycled sari clothes are hybrid products, constructed images that do not represent any particular society but are a material objectification of the perceived relationship between elements of two societies at a particular time (Miller 1991). The clothing is neither Indian nor Western in origin. In this case, the mix is of Western costume and Indian fabrics, permitting wearers to incorporate exotica into the style they consider fashionable without reference to the ways in which the cloth is perceived or used in India. Miller draws on Said's description (*Orientalism,* 1978) of how the West constructs the Orient as an oppositional image and then expects it to conform to that image. But this process can go both ways. The "oriental"-style chintzes that were developed in India for the British market in the seventeenth century, Miller says, "may be said therefore to be an objectification in material form of the concept of orientalism" (1991, 60).

Whether in a shop in Pushkar, a stall at a music festival in Europe, or a boutique in London, each recycled object is of a style that has been replicated and displayed *en masse* rather than being presented as somehow unique. In fact, this replication is itself an important aspect of creating authenticity, as Steiner has remarked. For him, this seriality is a "curious form of authenticity born in the shadow of mass-produced images and texts. . . . Just as in Benjamin's model of mechanical reproduction, 'the copy becomes associated with the true and the original with the false'" (Steiner 1999, 95).

Kingston shows how authenticity is a magical effect, similar to the "technology of enchantment" (Gell 1992b), relating specifically to its invisible technical origins (Kingston 1999). Therefore the originator of an item must be out of reach for that item to be "authentic." It is because they understand that authorship must be obscured (through the chain of connections inherent in the objects' production, beginning with the original owners' dealings with the *bartanwale*) that entrepreneurs are able to market their products successfully.

Recycling or Reincarnation: The Elite Indian Market

During my fieldwork, women in the Progressive usually claimed that Indians would never wear second-hand clothes from the market unless they were too poor to afford new ones. Middle- and upper-middle-class housewives and professionals alike denied that it could ever happen, and there was a general consensus that middle-class Indians would always buy new. The fact that there was a surplus of used saris was not surprising to them; it would obviously happen, because clothing wears out, fashions change, and gifting is so ubiquitous. But for them the old clothes can only ever travel down the social scale from richer to poorer. The exceptions were clothes that were inherited, swapped, or handed down within families, which usually traveled down the social hierarchy from the old to the young. The Western middle-class practice of shopping for second-hand clothing as a means of obtaining unusual fashions (McRobbie 1989; Gregson and Crewe 2003) was deemed inconceivable by middle-class Indians.

However, those lower down the social scale sometimes indicated that if middle-class people thought they could get away with it, they would. Waghri *pheriwale* in Ahmedabad spoke of selling to dealers who quietly sold clothing to the "Marutis," (the upwardly mobile middle classes whose nickname derives from the small Japanese cars they now drive by the million), but I never met anyone who would admit to buying clothing in this way. The exception is the upper classes in Delhi, who might purchase unusual "antique" saris to wear to special functions. It was therefore a great surprise to see an article in the fashion pages of the *Hindustan Times* (March 24, 2000) entitled "Antique Chic." Subtitled "The Millennial Woman and Her Designing Man," it featured the upmarket designs of Sonam Dubal, who uses "recycled" saris to make his clothes.

Sonam Dubal was then in his late thirties and had graduated from the first course in fashion design at the National Institute of Fashion and Technology (NIFT) in Delhi in 1988. He went on to work under Rohit Khosla, one of India's leading international designers, before establishing his own business. Sonam felt that he was therefore heavily influenced by the older generation of Indian designers, whom he conceived of as primarily textile artists concentrating on designing with fabrics, but following conventional garment forms. Under the label Sanskar, he designed upmarket Punjabi suits for sale to the Delhi elite, mainly retailing through a boutique called Ogaan in the stylish Hauz Khas district. The Sanskar range is made from natural fabrics with understated decoration using traditional craft techniques of embroidery, *qalamkari* (drawing), and printing, but Ogaan's clothing tends to be more innovative than that found in neighboring boutiques, and it therefore appeals to a younger, often well-known, clientele. Indian celebrities such as the writer Arundhati Roy wear Ogaan's clothing abroad to promote contemporary Indian fashion, moving away from the more expensive establishment designers such as Rohit Bal.

At the end of the 1990s, Ogaan opened a new boutique in the upmarket Santushi Complex, where retailers focus on high-quality Indian products for both the Indian elite and the expatriate community. They offer "traditional" garments, such as hand-dyed, handwoven scarves; items by contemporary designers who use Western design forms to inspire their Indian ranges; and clothes by new designers catering to the young people who favor Western and cross-over clothing with strong Indian influences. Ogaan's branch in Santushi aims itself toward the latter group, showcasing six or seven designers who mix elements of the latest European fashions, Indian fabrics, and original decorative schemes to appeal to the party set. The young designers are also from elite families; some are NIFT graduates, they often travel abroad, and they are highly aware of international fashion. The rise of Indian influence in the top end of designer fashion in Europe has enabled these designers to experiment with locally available fabrics and skills that are in strong demand globally amongst the trend-setters, although the designers are not yet taken as seriously by the big names in the business as they would like. The Delhi market is small and the fashion cycle is extremely short: to keep the boutique in business, young designers are expected to come up with new lines every two months, a tough but stimulating challenge.

Sonam designs a second range of cutting-edge contemporary fashion for Ogaan, and his trademark is the reuse of brocade saris. The clothes he makes are remarkably similar to those made by Tommy Frog. "Butterfly" blouses are halter-neck, backless silk tops with shoestring laces; long pencil skirts are slit to the thigh to reveal contrasting linings; three-quarter-length trousers are worn with matching "aprons" following the 1999–2000 European fashion. A transparent black chiffon blouse, with sari borders at the neck and cuffs, is designed to be worn over a recycled brocade brassiere. For more formal wear, the traditional long Indian *sherwani* coat is made with matching silk straight-legged trousers; waistcoats and jackets are sold as separates. Sonam uses a mixture of rich Banarasi brocades, sari borders, plain chiffons, and sand-washed silks, the latter creating harmony within the range while the decorative pieces of sari provide richly varied colors set within gold and silver tracery. Tops and skirts are plain with sari borders attached along the bottom, while the coats and jackets are made from the most expensive all-over brocades to be found.

This clothing is of a much higher quality than Tommy Frog's, with more sophisticated tailoring, better finishing, and infinitely more unusual fabrics; Sonam always uses silk, never synthetics. He obtains his saris from a "girl near Janpath" who knows what he is looking for and keeps pieces for him. She is a relatively cheap source of fabric for his designs, yet the saris she sells him are still more expensive than those for the export market: he often pays several hundred rupees for each one. The all-over silk brocades he chooses are skillfully woven in bold, bright color schemes and have a high gold content, while his sari borders are of a much finer quality than those in Camden Market, balancing the subtleties of color with delicately woven motifs. He is selling his clothes to an elite market with an intimate knowledge of Indian silks and brocades, and therefore uses fabric of a quality that his customers habitually wear. The saris he buys do not have their values inverted; his business involves no ironic picking up of cheap fabrics, of the sort that would be bought by the lower classes, to be remade into a quality product.[7]

Sonam himself has a deep understanding of the history of fabrics and designs, picking out expensive brocades that were fashionable in the 1950s, '60s, or '70s with designs no longer made today. In this he is unlike many of the Western entrepreneurs discussed above, who may not need to grasp the subtle social differences between an expensive, hand-embroidered chiffon and a cheaper mass-produced one, or choose a Banarasi silk brocade over a synthetic imitation, to satisfy their less knowledgable Western clientele. The significant differences in their clothing's value are of minimal importance once it has been shipped abroad. In Delhi, Sonam's careful selections enable another, deeper layer of understanding to operate between his clothes and their Indian buyers. Although fashion eras are not important to him *per se* in his use of the fabric, he enjoys the color contrasts used in earlier decades and likes to evoke certain design trends. Brocades have always been woven in designs that complement contemporary jewelry styles and the wider material environment of carvings, paintings, window tracery, and interior design. Sonam sees the cloth as a design whole within its surroundings, wrapping the body and linking it to the environment. His mother, a Sikkimese Tibetan aristocrat, was a socialite who loved to dress up and was herself known for mixing heavy Chinese brocades with the traditional Tibetan dress form, the *chuba*. He remembers her skills in blending her own family background with the Maharashtrian Indian elite into which she had married, creating an original and stylish look.

Some brocade designs date back centuries, to Mughal influence, and others bear even older Hindu motifs; the recurrence of such motifs over time allows them to appear quintessentially Indian, drawing as they do on a millennia-long heritage. Simultaneously, the brocades nostalgically remind women of the last time their motifs and colors were fashionable; perhaps they were worn by their mothers and grandmothers. The appeal of such fabrics may be likened to the recent vogue in the West for 1970s textile designs, with their color or form updated for the new season, which recall childhood for a new generation of adult consumers. For Sonam, the recycling of images, colors, and design motifs into unusual combinations and startling new forms allows him to represent clothing as at once new and yet familiar, to allow his customers to artfully experiment with new forms using classic cloth nuanced with layers of time.

For example, a man's coat is made from roughly woven, bright saffron cotton of the type worn by holy men and pilgrims, but its lining is a bottle-green silk brocade with Chinese dragon motifs. As the elite girls don their clothes for their exclusive parties, they are putting on a performance that asserts their claim to preeminence, reassembling elements of fabrics, colors, and motifs into new forms, and reconstructing themselves in the process. His collection becomes a sophisticated articulation of being young, wealthy, and Indian in a global environment, of being part of the elite, laying claim to the past and taking hold of a future that extends well beyond India.

In the more conservative Hauz Khas branch of Ogaan, Sonam has also sold a range of traditional Punjabi suits made from used silk saris. In particular, he utilized plain silk saris that had a "shredded" look, that is, old saris that had been worn or folded until creases appeared, and in which the weft had started to slip. He made them up as linings or backed them with a bright lining in a contrasting color that showed through the weft. Sonam tells a story of how one woman saw them and walked out in disgust, complaining to the boutique owner about the range—for it is exactly the type of worn-out old sari that women in this class would not normally be seen dead in, and that finds its way to the *bartanwale* in the first place. However, after talking to her friends, she must have realized how unfashionable her moral stance was becoming, for she was back soon afterward hoping to buy one, and was bitterly disappointed to find they had all sold out. Sonam also knew that his consumers would not "blindly turn to the West and accept new materials and new fashions," but needed to feel that the ideas were part of an Indian heritage. The designers and retailers have skillfully managed to create a new fashion by reversing a deep-seated anxiety about used cloth, turning it into a desirable commodity amongst the elite.

Overcoming the usual reluctance on the part of the Indian upper and middle classes to buy second-hand clothing cannot, therefore, be taken for granted. Younger buyers are certainly happy to buy the new styles and seem less concerned, but Sonam still markets these fabrics as "antique silk," cashing in on the principle that "old is gold." Antique silk saris, shawls, and textiles in general have long been highly prized amongst the upper classes, as they were during the colonial period and are still by contemporary collectors and dealers. Yet, although some antique pieces have always been cut up and reused when worn or damaged, significant value is also held within the form of a whole sari or shawl. This usually, but not always, prevents knowledgeable people from cutting up potentially valuable textiles. In addition, older women know well that these pieces are not really "antique," and therefore they need extra reassurance that they may cross uncertain moral boundaries.[8] This is partly provided by the subtle message inherent in the quality of the silk itself, which they feel must have belonged to "people like us."

To allow these women to bypass the stigma of wearing second-hand clothing and justify the use of pieces of old brocade, Sonam appeals to two factors that can counter the fear that material from unknown sources will bring social disgrace and ritual pollution. One is the Indian consumer's growing awareness of the Western view of recycling as an environmentally friendly practice, and the Indian precedent for this view in the practice of thrift. The other is the argument that recycling as an ancient process embedded within Eastern philosophy. All his garments have this label attached:

> With the dawn of the new millennium global awareness becomes:—
> Fabric used is Antique Silk infused with the concept of recycling so
> flaws in fabric weave or texture are to be considered NATURAL as to
> any woven traditional fabric aged through the process of time.

In an early edition of the new Indian *Elle Magazine,* launched in 1997, Sonam's label was featured through an image with global appeal. A pale-skinned, middle-aged woman, who might have been either an Indian or a Westerner, wore a silk "Butterfly" blouse. Her eyes were closed, and she had four or five acupuncture needles in her face. The caption read, "We are all spirits here—visiting—moving on—to different times. Reincarnation is the theme Sanskar is based on." By the mid-2000s, after many years of networking, Sonam's clothing was on sale in boutiques in Delhi, Goa, Mumbai, Kolkata, and Bangalore in India, as well as in New York and Europe, and his lines were regularly shown in the new Asian fashion shows such as India Fashion Week and Singapore Fashion Week (plate

16). Although he has branched out into various ranges, the mixing of old and new fabrics remains central to his design repertoire. He sees his clothing as a combination of Indian and Oriental styling which works across boundaries, fitting into an Asian cosmopolitan trend with broad roots that can then be marketed to the West.[9]

The famous couplet from the *Bhagavad Gita* (2:22) quoted in chapter 1 discusses the transient nature of material form, likening the way that the soul, *atman,* periodically casts off the body to the way a man casts off his old coat and turns to a new one. Sonam also turned to Indian philosophy to explain to me the creative process behind his clothing, revealing that he found Buddhism both personally inspiring and useful as a marketing tool. In the article in the *Hindustan Times,* he is pictured in a yogic meditation pose, *padmasana.* As a Buddhist, he meditates daily upon the five elements in nature, water, fire, earth, air, and the ether, and on how their individual forms break down as the elements dissolve into one, pure formlessness. The Hindu equivalent is the concept of *maya,* the illusion of the material world, which ultimately must be broken down through pure perception. Sonam also draws on notions of cyclicity, destruction, and regeneration. He cuts out elements of the cloth, such as pieces of gold and silver, colors, and motifs, and builds them up again into new forms. He describes the process as one of renewal, of creating new life, in which fashion and clothing are essential means of wrapping and re-forming the body within its environment. Despite the obvious unspoken business advantages of the current global fashion for "Indian" clothing, and the easy availability of the cheap old saris, it is this process of renewal that Sonam chooses to focus on when explaining his success—he believes it will support his creativity for a long time.

The use of old clothing in the marketplace reveals the material framing of value judgments of the fibers and cloth of which it is constituted. Whilst old rags depend upon the softness and absorbency of natural cotton, clothing sold to the poor tends to be made from durable synthetics or designs which can be easily remade for regional markets. But as value lies within the fabric itself, it becomes a resource for the creation of new hybrid products, utilizing the silk, designs, and "Indianness" of old saris to construct new objects.

Thus the singular garment in the wardrobe creates a new category of hybrid seriality, constituting and making visible a new set of relationships through the destruction of the old. The marketplaces are overflowing with used cloth, stripped of its role as a locus of identity. Yet, although the origins of such cloth may not be particularly problematic for many non-Indian consumers, more subtle and complex manipulations of value are required for the elite home market. For these products to be accepted, a sleight of hand must transform the used garment into an antique. The saris have their origins concealed and are made to appeal to buyers on the grounds of being authentically linked to an older period, suitable for a higher class, correlating with Hindu cosmological values of cyclicity and the Western concept of recycling as an environmentally responsible practice. The politics and practicalities of hidden transformations are essential for the translation between regimes of value and the creation of new modalities.

FIGURE 7. Second-hand clothing is washed and hung out to dry on the barriers along the main road. PHOTO COURTESY OF TIM MITCHELL

7 VALUE AND POTENTIAL

 The most powerful and evocative images arising from this research are of mountains of old "stuff": piles of shiny silk saris and cotton suits escaping from bulging metal wardrobes, suitcases with strained locks and trunks with broken hinges. As clothing is pushed across the threshold of the domestic sphere and out into the wider world, this flow of used, second-hand, polluting, and inauspicious substance manages to remain invisible to the eyes of most middle-class Indian consumers. The baskets of pots that the Waghri women balance on their heads metamorphose into unwieldy bundles of used garments. Every day, the Ghora Mandi market is filled in the dark early hours of the morning with heaps of unwanted cloth, the most valuable glittering under the nearby streetlights. As the market empties of goods, traveling salesmen leave the city to sell the old clothing in the outlying towns and villages, while local stockroom shelves are replenished with piles of silk fabric. Families are busy putting together consignments of furnishings and clothing for export overseas, where they populate the festivals, shops, and retail catalogues which feed the Western imagination and fuel the desire for the exotic.

The journey through the landscape of old clothing exposes an ecology of accumulation and overflow that overwhelms the discursive narratives of containment and control. The anthropological focus on cloth in India has until now remained fixed upon how clothes mark identity through their production and consumption and the strategies whereby their potential can be manipulated. However, this study has crossed an invisible barrier and looked at the hidden processes and alternative trajectories of cast-offs and rejects. Instead of regarding the West as the source of waste and the Rest as its dumping ground, this book has focused upon strategies of dealing with increasing amounts of unwanted surplus in a developing region.

Ridding

Essential to the understanding of this reversal of clothing's meaning is the perception of the materiality of cloth as a resource to be utilized in the unmaking and remaking of persons and identities. The technologies of making thread, weaving, and gifting and wearing cloth provide possibly universal metaphors for the creation of the cosmic order, social relations in the human world, and the interaction of these two levels. In classical Indian tradition, the universe is conceived of as a fabric woven by the gods. "The cosmos, the ordered universe, is one continuous fabric with its warp and woof making a grid pattern . . . hence the importance of wholeness . . . of the uncut garment, like the sari or *dhoti*" (Kramrisch 1989, 79).

Missing from these more usual insights are those equally powerful images suggested by the fragility, impermanence, and susceptibility to decay inherent in the fibers and technologies of making cloth itself. The management of wardrobes as integral parts of the relational self, overflowing with gifts and hand-me-downs, demonstrates how its unraveling, fraying, and disintegration allow cloth to operate metaphorically as the ideal sacrificial victim, enabling the reordering of the social world through its rejection, destruction, and subsequent recycling. It is through daily practices of caring for clothing, folding and storing, washing and ironing, that clothes are reappraised and reclassified, until they are deemed no longer fit to remain in the wardrobe.

In treating objects as persons, Gell acknowledges his debt to Mauss's theory of exchange and the substitutability of persons and things (Gell 1998, 9; Mauss 1954). Clothes are considered "skins," the exuviae of a person; they do not "stand in" for a person metonymically, but are physically detached fragments of someone's "distributed personhood" (Gell 1998, 104). Clothing is clearly worn on the body and is believed to have influence over it by virtue of this proximity. Yet as an image, clothing also shares properties with its owner through representation. Clothes are the archetypal "skins," comprising both images and fibrous layers shed from the surface of the body. People change, their relationships evolve, their status

may wax and wane, they grow old. Clothing constructs images and representations on the surface of the body that are crucial for identity but that
need constant updating.[1] The notion of internal growth is also central to
Campbell's argument that fashion and the desire for the new provide the
cyclicity of fantasy, fulfillment, and disillusionment in the development of
the person (Campbell 1992).

The disposal of unwanted clothing allows for renewal, the shedding of
connections, the breaking of social relations made by its giving and wearing. Cloth can do this in a way that other, perhaps more durable, things
cannot. Items of clothing become disembodied cloth through being rid of
the bodily substance once held within them and their subsequent cutting
and remaking into new objects.

Riddance and shedding beyond the domestic realm are less often spoken about in middle-class Indian households. The breaking of social relations is overlaid with the language of continuity and conservation. Almost
secret in some families, such strategies allow clothes to move between regimes of value and trash to turn into treasure. Cloth makes change apparent through its materiality and the manipulation of its various properties.
These properties look back to the past, constitute the present, and help to
create the future, using representations of continuity and discontinuity.
The wardrobe is the nodal point in this flow of clothing: some clothes re-
side in it for a lifetime, whilst others are acquired and disposed of almost
immediately. Temporal and spatial relationships connect people, places,
and things; once these relationships are ruptured, cloth becomes a means
of interweaving new pathways and ways of being. The continual shedding of clothing throughout the life cycle allows for smaller, incremental
change within the larger arcs of the life cycle.

Recycling

Clothes are not passively recontextualized through their transformations
and travels, but constitute a resource in the mutual reconfiguration of
person and object and in the individual's shaping within a social setting.
Küchler and Were speak of the "strategic transformation of clothing
through the shifting of motifs onto new surfaces, the decomposition and
reassemblage of materials, and the shift in associations bound up with

clothing through performance and art" (Küchler and Were 2005, xxii). Benjamin envisaged translation as the material act of making connections visible in surfaces that resemble each other; such surfaces are corporeal, they resemble the body's surface, whether they are on or off the body (Benjamin 1992a).[2]

The global travels of cut-up and refashioned decorative silk saris suggest that perceptions of the generic "orientalism" of sari designs allow for the interplay of the exotic and the familiar on the body. Any claim to authenticity is indeed linked to a distant, ontologically invisible domain; the original owners of the saris are unknown, and their dreams, thoughts, and lives will never be known; those cloths are stripped bare of their former meanings. The mass of fabric overflowing in the market testifies to the ready availability of such means for the reconfiguration of new identities, in sharp contrast to those contested badges of identity that are viewed as a scarce resource (Harrison 1999).

My two final accounts of recycling are both examples of the total effacing of clothing's origin and ethnicity, which might be seen as the last marker of a previous life. Complementary yet opposing transformations of cloth, the recycling of saris in the UK by British Asians and the shredding of Western clothing in India for the shoddy (remanufactured wool) trade, both radically remove all traces of belonging. As Sital Punja, founder of Sari UK, said, "If you take out the ethnicity of a sari you open it up for someone else, someone who isn't Indian or a hippie or has older ideas of recycling as boring and worthy . . . you make a space for something new."

Sital founded Sari UK in late 2004, and has cleverly manipulated both Indian and Western ideas bound up with old saris in order to produce a recycled product and donate part of her profits to children's charities. Preparing for a trip back to India, she was weighed down by the old saris of well-meaning relatives in London and instructed to take them back home to north India and distribute them to various cousins. This is still a common way for first-generation British Asians to recycle their saris; women who have emigrated feel good for having given to people they believe to be in need, and they can feel guilty if they do not do so. These days the recipients of such hand-me-downs, although polite, are not always so grateful; but there is no significant market in the UK for saris that have been donated to charity.[3] Sital perceived the mismatch between the piles of unsuitable old saris and the unmet need for practical help. So she founded a business that would collect old saris in the UK using networks of temples and Saturday clubs and hire young freelance designers from around the world to turn them into high-end fashion items, selling for hundreds of pounds in London's top boutiques. By giving part of the money to charities operating in India, she satisfies the donors' desire to do good. The saris are totally removed from their cultural context; they have slipped through the lines that tie Sonam Dubal's clothing back to the Indian elite and are no longer anchored in a recycled mix of pure silks, traditional patterns, and Western designs that just push the limits of Indian respectability.

Sari UK plans to develop social enterprise initiatives with marginal British Asian women and bring them into the workforce, and to develop an eco-design network. The old silk saris are now implicated in a growing nexus of concerns that attempt to simultaneously satisfy the need to re-use and recycle, to feel good whilst doing so, to act in an environmentally aware manner, to create a sustainable business and generate profits, and to become instrumental in creating models of social enterprise and inclusion that can be translated across cultures and used to help immigrant groups across the country. The saris themselves are materially translated into high-end Western fashion, in turn valorizing the myriad of linked transformations that they have made possible.

The shoddy trade presents a strong contrast to this reuse of saris within the UK. Cast-off Western sweaters and coats are turned into new Indian blankets via the shoddy processing mills in Panipat, Haryana (Norris 2005b; Norris 2006). Here, old woolens are sorted into color families and stripped of their fastenings, linings, and labels before being industrially shredded and pulped. Then the fibers are teased out, carded, and spun into brightly colored new yarn, ready for reuse. All the associations of the clothes have been completely effaced, including the garment labels that bore witness to their design heritage and place of manufacture, the bodily traces of former occupants, and the style and form of the garments themselves. All that remains is a heap of fibers twisted into a new raw material. These threads are rewoven into shawls for the poor in hilly regions, thick fluffy blankets for the lower middle classes, and lengths of suiting for school uniforms, men's clothes, and even the garment export trade. Even more radically than the reused saris, these clothes have been completely de-ethnicized; all that remains of their former lives are the traces of grime and dirt held within the fibers that are never washed away. Instead, new motifs are applied to the fibers; blankets are decorated with typical Indo-Persian "vase of flowers" designs, while suiting is woven in contrasting checks and tartans. The new cloth and clothing is reoriented toward the new potential consumer, inserting the fabric into a new nexus of consumption, appropriation, and agency.

This study confirms that the value of clothing does not lie solely in its meaning as embodied material culture. It is not necessarily the shape, style, or cut of clothing that defines its value, nor the social relations that it encapsulates. Instead, a value inherent in the discarded "stuff" that is shed from the body awaits discovery, and that "stuff" can be a resource to be remade into new things.

Keane suggests that by treating objects as merely a sign or symbol of something other than themselves, we lose sight of their actions, consequences, and possibilities within a world of causality (Keane 2005). He refers to Nancy Munn's "qualisigns," which are certain sensuous qualities of objects that have a privileged role within a system of value, such as "lightness" or "fluidity" (Munn 1983; Munn 1986, 16–18). These sensual qualities have significance beyond any particular manifestation of them, i.e., any objectification of them in material form. This leads to an understanding of icons as unrealized potential, a means of discovery through the effects and suggestions of material qualities. The realization of one potentiality depends in turn upon the subordination of another, and these realizations often depend upon the contextual and contingent "bundling" (as Keane describes it) of certain signs or qualisigns. Clothing presents a prime example of the bundling together of various material qualities that have contextual value.

In the disposal and reuse of clothing in India, the concept of the unbundling and disaggregation of properties is of crucial importance. Sensory properties such as the smell of an old garment, the softness or starchiness of a fabric, the bright or faded color of clothing, its design and patterning, the style of its cut, and the real or imagined history of its use and reuse can all be foregrounded or pushed into the background, and recombined in as many ways as can be reimagined and recrafted. An older woman in Delhi will ask a tailor to cut up a favorite sari to make a granddaughter's party dress, at once sad to see it destroyed but hopeful that in another form it will be useful and cherished once more by a younger loved one. She will fondly tell the girl how her husband presented it to her years before, when her first child was born, and by wrapping the child in that cloth she brings her within the folds of her family, tying her a little closer for all to see. A tailor in Pushkar will pick up a similarly old silk sari with brilliant designs from a bundle of hundreds he has bought from a Waghri seller, oblivious to the fact that it was once part of a trousseau and that the large stain that makes it unwearable was caused by the clumsy hand of a child. He will carefully construct from it a halter-neck top for a young Western woman, preserving the patterns and gold *buttis* that conjure up images of India and Indian fabrics, and creating a potential souvenir, a hybrid garment for the here and now, something that is of India but not, perhaps, quite "Indian."

As Thomas suggests in his discussion of the introduction of cloth and Christianity in Polynesia, it may be more useful to understand culturally mixed forms not simply as hybrid objects or expressions of hybrid identities, but as techniques of individual and social transformation (Thomas 2000, 199). Hybrid textiles make possible new forms of embodiment: the travelers moving through Pushkar are able to act and perform in novel ways through the clothing which they can have designed or buy ready-made. It enables them to be in the Rajasthani desert, wrapped in gold, silver, and silk, choosing their own sartorial terms of engagement with Indian dress, which simultaneously allows them to be part of a well-defined group of similar Westerners.

The complex strategizing of individuals in choosing how to rid themselves of clothing, and of entrepreneurs in deciding upon the best use of the old clothes at their disposal, demonstrates how different factors become important as value is continuously constructed through the exchange process. This complements Munn's proposition of the reciprocal construction of value (here in relation to the *kula* exchange in Gawa): "although men appear to be the agents in defining shell value, in fact, without shells, men cannot define their own value; in this respect, men and shells are reciprocally agents of each other's value definition" (Munn 1983, 283).

Strategies of divestment and investment connect the local to the global through the materiality of cloth and the material properties of fiber, weave, color, and shine. They reveal a landscape of old clothing whose vibrant colors, patterns, and textures are reconfigured to decorate and clothe new bodies and homes, but whose origins must remain invisible. The stuff of cloth, its malleable, mediating materiality, is the shape-shifting substance that takes on form to create such networks; it joins and separates, inserting itself between persons and things and enveloping a new wholeness before disappearing once more.

 NOTES

1. RECYCLING INDIAN CLOTHING

1. Between 1980 and 1989 India witnessed a consumer revolution, with a 47.5 percent increase in consumption expenditure and more than 100 million people with sizeable disposable incomes (Dubey 1992).

2. Interview with Santosh Desai, of the advertising agency Ogilvy and Mather, New Delhi, May 30, 2000.

3. Chakravarty and Gooptu explain that

> the expansion and growing prosperity of the middle classes occurred within an interlocking context of the opening up of markets, extension of investment and credit facilities, increasing public pay packets, tax cuts, better entrepreneurial opportunities through privatisation and the removal of bureaucratic controls. [This led to] an increase in purchasing power, an expansion of consumer habits and a boom in consumer products. (Chakravarty and Gooptu 2000, 91)

4. Barthes' analysis of the Western "fashion system" reveals how it relies upon built-in obsolescence and high turnover, which may be increasingly true of middle-class Indian fashion (Barthes 1985).

5. In particular, the predominance of the sari and *dhoti* (loincloth) in Indian clothing demonstrates this close relationship with cloth, which appears to have been lost in the daily lives of many Western consumers. Even where wearing stitched clothing is the norm, the vast majority of women still buy up lengths of cloth and take them to the tailor to be stitched rather than buying ready-mades. Desai states that "Indians buy far less clothing than Americans, but far more cloth . . . services are cheaper than finished goods, so people buy things in a less finished form and prepare them for use themselves or use hired help" (Desai 1999).

6. Greener Style, "Recycled Silk and Sari Bags," http://www.greenerstyle.co.uk/bags-recycled-sari-bags-c-54_81.html (accessed August 12, 2008).

7. Namaste UK, "Recycled Silk Sari Skirts, Dresses and Trousers," http://www.namaste-uk.com/section.php/1337/1 (accessed August 12, 2008).

8. Sari UK is one exception, to which I return in the last chapter.

9. The importance of European clothing, some of it second-hand, as a vehicle for change in colonial Brazzaville has been discussed by Martin (1994). Academic research on clothing and colonial contacts with indigenous populations tends to focus on the adoption of new European-style clothing or the role of hand-me-downs in domestic contexts, while little work appears to have been done on the history of the international trade in old clothing. Undoubtedly other colonies were also destinations for commercially traded cast-offs, both of reusable clothing

and of rags destined for the paper industry, but there was also a flourishing market within Europe. Lemire (1988, 1997) shows that in preindustrial Britain, before the advent of mass production, second-hand clothing was viewed as legitimate wear for those from the poorest ranks up through the middle classes.

2. FIELDWORK CONTEXTS

1. My neighbors spoke Hindi, and often had another Indian language as a mother tongue, but most residents also spoke good English, and this was the language we would usually use together.

2. In 1999–2000 the Maruti was still the most readily available new car in India, although the situation has changed dramatically since with many models for sale.

3. Of course, these views were not espoused by all of the Progressive's residents, and many felt very differently about community cohesion and their experiences of living there.

4. This was the case in the modern housing developments across Trans-Yamuna during my fieldwork. However, trials of domestic rubbish sorting had been started in various locations in central and south Delhi, and subsequently more projects have been undertaken to develop waste collection and processing systems in the city. None include clothing *per se*. Chaturvedi's earlier study of materials recycling in Delhi (1994) also found no textile recycling.

5. Hereafter referred to simply as "suits."

6. Middle-class urban Hindu women are still expected to wear saris at any social or religious function, and they remain the daily choice of many, whether they are going out to work or running the home. Suits are also popular everyday wear, whereas Western clothing is generally worn by men, children, and young women, but rarely by married women. Teenage girls and college students wear long cotton skirts or jeans and cotton tops, moving to suits as they grow up and start working; once married they are normally expected to begin wearing saris, at least for a while.

7. The fall is a strip of matching cotton about six inches deep that lines the lower hem of the sari and protects it where it touches the ground.

8. The social and cultural life of Delhi is most active in the winter, with concerts, plays, and dance performances as well as family celebrations.

9. I was also interested in men's views on the value of their clothing, and I spoke to men about their own attitudes in several instances, but they would always refer me to their wives for answers to more detailed questions about how old clothing was valued, as it is they who "sort out that side of things." The management of wardrobes remains firmly within the female domestic economy.

10. Such resourcefulness has been recognized in the West for its aesthetic creativity and as the inherent materialization of the entrepreneurial spirit. Two exhibitions have helped give "recycled art" an international status: one curated by the Pitt Rivers Museum, Oxford (Coote, Morton, and Nicholson 2000), the other by the Museum of International Folk Art, Santa Fe (Cerny and Seriff 1996).

11. During 1999–2000, Rs 100 was equal to about £1.50 or US$2.30.

12. The terms *chindi* and *katran* are part of an elaborate system of classification, by size and fiber content, of waste products from the garment manufacturing

industry. Similarly, the earlier stages of cotton production create waste products which are either recycled back into mainstream production or sold on to related companies as a cheap raw material. Scraps of cloth can be used in the papermaking industry; for example, in Ahmedabad a factory established by the Gandhi Ashram to promote recycling and self-sufficiency produces handmade paper. Lemire (1988) notes that used clothing was sold for papermaking in preindustrial England, and a dearth of rags in the mid-nineteenth century resulted in the importing of Indian rags (Steedman 1998).

13. There are a hundred paise in a rupee.

14. *Chindi* durries are now made for export in large quantities.

15. It is highly likely that some middle-class informants did in fact buy imported used clothing for themselves in secret; the belief that such behavior existed was firmly established and some women claimed that although they themselves would not dream of it, they suspected others of doing so. One woman did buy her live-in maid used sweaters in winter, and this could be explained as sensible, thrifty behavior. Yet, unknown to these women, some of the better-quality "new" products on sale are, paradoxically, profoundly old and truly dirty. For the very fibers of many apparently new blankets and woolen goods are, in fact, recycled from imported woolen clothing, a fact which is completely concealed (Norris 2005b).

16. In fact, some Tibetan families in Dharamsala make large amounts of money by selling the large amounts of winter clothing given to them by departing foreigners (Audrey Prost, personal communication), but overall this is an insignificant source of trade.

17. See Hansen 2004 for a discussion of the controversies over this. Placing the trade in its international context, she addresses the issue of whether such clothing benefits the importing economy or threatens indigenous production. Haggblade surveyed the question in the context of Rwanda (1990), and Baden and Barber reported on the issue in Senegal in a report for Oxfam (2005). All conclude that overall the trade is likely to benefit both small traders and consumers by providing the cheapest clothing, and has less impact on the production and consumption of local garments than might be assumed, largely because of competition from East Asia.

3. LOOKING THROUGH THE WARDROBE

1. More recent research into wardrobes in the UK has also revealed the complex, multi-layered, and often ambiguous relationships between women and their clothing, and how many categories of clothing are never worn (Woodward 2007). This has now been followed up by a study of how people dispose of unwanted garments (Fisher et al. 2008).

2. Bayly discusses how the importance of clothing's source and of the occasions attached to it was historically magnified by the belief in the capacity of place and event to affect the materiality of the cloth itself (Bayly 1986). This magnification is illustrated by Cohn, who cites passages from Abu'l Fazl's description of Mughal rule under Akbar. The vast textile tributes pouring into the capital were managed by documenting the days of the week and month each item arrived, as well as its price, color, and weight; clothing was ranked accordingly (Cohn 1989, 315–16).

3. Saris for a wedding trousseau, gifts of money, donations to temples, and all other gifts that can be numbered are often given in multiples of ten "plus one," making an auspicious number.

4. A love marriage is distinguished from one arranged by the parents of the bride and groom.

5. The Hindi term *almari* derives from the Portuguese *armario,* itself from the Latin *armarium,* a chest or safe.

6. As Cwerner shows, the wardrobe is not only an object in its own right, but also "commands a set of distinctive and identifiable *spatial practices:* forms of structuring, delimiting, and organizing clothes, as well as the social meanings and identities articulated by these forms . . . the wardrobe articulates, both spatially and temporally, a set of material and symbolic practices that are fundamental for the constitution of selfhood, identity and well-being" (Cwerner 2001, 80).

7. Such cupboards cost approximately Rs 20,000 in the year 2000.

8. In Mehta and Belk's study of favorite possessions amongst urban Indians, the Godrej cupboard was frequently cited by women, second only to family shrines. They quote a Mrs. Rao, aged fifty-five, as saying, "Recently . . . I lost the key to the Godrej. I cried for two days. We then had to get the locksmith to open [it]. . . . It was the most horrible feeling. Some outsider was going to lay his hand on my cupboard. The feeling I got was similar to [the] feeling one gets when someone hurts your children. This Godrej has been with us since we were married. It has travelled everywhere with us. Now our children have gone and settled elsewhere; only the Godrej has remained with us" (Mehta and Belk 1991, 404).

9. The *mekhlar* is a tube of cloth worn as a skirt, with three pleats in front. The *chaddar* is a wrap about two meters long which comes from the back over the left shoulder in pleats and wraps across the front of the body: to the right side, then around the back, than back to the front again, where one corner is tucked into the waistband of the skirt. Both are heavily decorated with typical Assamese motifs, and a matching blouse would be worn underneath.

10. Her ex-husband had recently asked her for all the clothing and jewelry back, but although she no longer wore the clothes (and had never worn some of them) she refused on principle, saying that he contributed nothing to his child's upbringing. She suspected it was his family wanting them back in order to provide gifts for another son's new wife. Under legislation concerning *stri dhan,* bride's wealth, she was entitled to keep anything given to her by her in-laws, but this is usually hard to enforce.

11. Assamese women are known for wearing more Western clothing than north or south Indians, but Bulbul still constantly risked censure for wearing her sleeveless sundresses and mid-calf-length skirts around the apartment complex or to answer her door at home. Some felt that her inappropriate clothes were in keeping with the unacceptability of her position as divorcee, single mother, and "mistress."

12. Clothing is most often formally acquired in Hindu culture in the gift-giving that surrounds the rituals conducted at various stages in the life cycle, the *samskaras.* Gifts of cloth, money, and gold (in the form of jewelry) are usually given or exchanged amongst the families concerned at marriage, and new cloth is given to the mother at the birth of a baby, at the child's weaning (around the fourth

month), and on the shaving of the child's head (tonsure). The whole family of a higher-caste boy receives new garments on the tying of the sacred thread.

13. See Corrigan 1989 for a comparative study of clothes swapping amongst UK families.

14. Children are not members of social categories: "the complete incorporation of the child in the social world only takes place with the initiation of the boy and marriage of the girl" (Das 1976, 257). Until this point, children do not receive new ritual clothing, and until recently this would *de facto* mean no new clothing at all in many poorer families.

15. In north Indian marriage patterns, the engagement usually involves the giving of gifts by the bride's family to the groom's. These are the first of many gifts that accompany the bride, known as *daan-dahej,* and the gift-giving continues through to the bride's establishment in her husband's household. In addition to providing clothing and jewelry for the bride herself, the bride's father should gift clothing to the groom's family and extended kin. For men, this may be a shirt and trousers or, less often, a *dhoti,* while customarily a sari and blouse piece are distributed to every kinswoman attending the wedding. The gifting of clothes is a very tangible element of support for a woman, signifying an undertaking to be a provider, as the men in a woman's family traditionally do (in north India) throughout her lifetime.

16. None of the women I spoke to conceived of such gifts as part of a woman's family's ongoing obligation to continue giving gifts to their daughter, her husband, and her in-laws after her marriage. However, the customary rationale for such practices may lie in this hypergamous flow of gifts, detailed for north India in the anthropological literature (Parry 1986; Sharma 1984; Vatuk 1975).

17. Rohini echoed the sentiment of many when she described the recent increase in money spent on display at weddings and parties, and commented on the increasing importance of having different stones and jewelry to match each colored silk.

18. *Shahtoosh* shawls are made from wool taken from the undercoat of a rare Himalayan antelope, the chiru. It is said that a *shahtoosh* (the Urdu word means "king of shawls") can be passed through a wedding ring. Because the antelope is an endangered species, the wool is now internationally banned, though still available illegally. Its price and rarity put it beyond the means of my neighbors in the Progressive, but I did see a few older examples when talking to wealthier women in south Delhi homes, including a whole sari.

19. The *set mundu* is a regional ensemble of two half-saris, usually in white and gold; one piece is worn wrapped around the lower half of the body with pleats at the front, and the other is secured at one end in the waistband in front, crossed over one shoulder, wound round to the front again, and tucked in.

20. Appadurai comments on the way the Indian middle classes used food to develop a national cuisine through the inclusion of regional specialties (Appadurai 1988), while Pinney draws an analogy between food and clothing: "edibility and wearability stand as parallel idioms of national integration" (Pinney 1993, 119). Banerjee and Miller note that it was Indira Gandhi who did the most to popularize the acquisition and wear of regional saris, as she would appear in public

wearing saris appropriate to the area she was touring (Banerjee and Miller 2003, 128–29).

21. The acceptability of Western styles is rapidly increasing in Delhi, especially in suburbs in the south of the city.

22. The same problems are of course commonplace in the Western context as well. Lurie observes that a gift of clothing "is a mixed blessing, for to wear clothes chosen by someone else is to accept and project the donor's image of you; in a sense, to become a ventriloquist's doll" (Lurie 1981, 22).

23. Vatuk described customary gifts given at weddings whose material might be of such poor quality it could not be worn, a practice that was openly acknowledged in the rural area where she did her fieldwork (Vatuk 1975, 163). Although exchanges of gifts between relatives should ensure that proper relations are being maintained, they also lead to numerous opportunities to display regard, favoritism, and affection, or alternatively disdain and scorn, any of which may result in unwearable gifts.

24. On a couple of occasions my own questioning prompted women to offer me old silk saris so that I could "make them up into something," and my rather sparsely decorated flat gained the odd sofa throw and cushion cover made from my neighbors' rejects. I thus became another conduit for getting rid of things.

25. This echoes Mines's findings in south India, that as people age they may feel less constrained by familial expectations and more able to create and pursue new options in life (Mines 1988).

4. LOVE AND PROTECTION

1. Parts of this first section were given as a paper at the 2003 meeting of the American Anthropological Association and subsequently published in the *Journal of Material Culture* (Norris 2004b).

2. Traditional Kumaoni clothing styles incorporate the need to use heavier woolen cloth and knitted garments in the cold winter months.

3. It is considered by some to be unlucky to stitch any diapers or wrappers until the baby is born, and the period of special danger also varies. Old cloth may thus be used for the first forty days, or for four months (until weaning).

4. Rohini explained that when someone with a black tongue says something bad will happen, it invariably does. For example, on a bright sunny day, her mother would tell her to take an umbrella, as it was going to rain, and it therefore usually did. See also Babb 1981.

5. See Corrigan 1989 on sharing clothing within families in the UK.

6. Her husband's relations perform a *puja* (ritual worship) for the dead *suhagin*s in the family whenever there is a family ceremony; they used to distribute a sari to each married woman, but now they just give each a blouse piece.

7. See Crooke 1896, 162–64, for nineteenth-century examples, and Mookerjee 1987, 11, for a discussion of the famous tree shrine at Chir-Ghat, Vrindavan, where Krishna stole the *gopis'* clothing.

8. Indian fabrics have long been prized trade goods (Barnes, Cohen, and Crill 2002; Guy 1998). The Newberry Collection of block-print Gujarati cloth excavated

at Fostat, Egypt, includes a piece of carefully patched cloth made up into a child's garment, which is now in the Ashmolean Museum (Barnes 1997). In early-eighteenth-century Britain, prohibited Indian chintz was being saved and patched together to make hangings and cushion covers (Irwin and Brett 1970).

9. Quilted war coats worn by Mughal noblemen contained from forty to sixty layers of padding within them for protection against swords. The war coat of Tipu Sultan, who fell in battle in 1799, contained, between layers of cotton and wool, several small pieces of seventeenth- and eighteenth-century embroidered pashmina shawls from Kashmir, some of which would have been over one hundred years old when used (Finch 1996). Several of the rare fragments were given to the Victoria and Albert Museum and are catalogued in Irwin 1973.

10. These robes spread across the Middle East and North Africa with Islam, and were brought into European culture by the crusaders as Harlequin costumes (Jasleen Dhamija, personal communication).

11. Having died out as a domestic craft practice by the early twentieth century, *kanthas* are now being made once more for the tourist trade and have become a contemporary art form.

12. According to Dutt, the traditional form was a mandala, often the *satadala padma,* or hundred-petaled lotus, surrounded by animals and humans (Dutt 1939, 460).

13. Strathern's work on composite personhood in Melanesia stresses the centrality of gifting to the creation of social relations, objects making such relations visible (Strathern and Strathern 1971; Strathern 1999). These relations are understood to "wither or flourish according to the properties seen to flow alongside them. The effectiveness of relationships thus depends on the form in which certain objects appear . . . it is the capacities in persons/relations which are reified" (Strathern 1999, 16).

14. The higher social classes' fear that their servants will wear clothing as fine as or finer than their own is found across many societies and periods and leads to various preventative strategies, including sumptuary laws and rapid changes in fashion (see Lemire 1988, 1991 for European examples).

15. Ginsburg notes that in mid- to late-nineteenth-century England vicarages were centers of distribution of old clothes, which were made suitable for the recipients by remaking them and stripping off all the trimmings (Ginsberg 1980, 129).

16. One Delhi-based NGO thought that middle-class Indians would soon be ready to give to a reputable door-to-door collections scheme, as the amount of unwanted wearable clothes had undoubtedly been increasing since economic liberalization. This is in fact the aim of the Goonj foundation (http://www.goonj .info).

17. Bataille goes on to accuse the modern bourgeoisie in the West, who now have wealth, of refusing to acknowledge their obligation of functional expenditure. "The representatives of the bourgeoisie have adopted an effaced manner; wealth is now displayed behind closed doors, in accordance with depressing and boring conventions. . . . [The bourgeoisie] has distinguished itself from the aristocracy through the fact that it has consented only *to spend for itself,* and within itself" (Bataille 1985, 124).

18. Such strategies recall the gastro-politics of cooking and eating scrutinized by Appadurai (1981); he links food transactions to Geertz's notion of "deep play" (1973).

5. SACRIFICE AND EXCHANGE

1. Although many professional women did go out to work, traveling by hired coaches or public transport, there was general disapproval of women "roaming" around the city, and I was often told that traveling alone, even by day, was unsafe and ill-advised. I had hoped that some of the well-educated young women in the Progressive who were not working would be willing to assist with my research, but of those who expressed an interest, in the end none were actually able to travel to any of the markets in which I worked, or could only do so if we were escorted by a suitable male, who was never actually available.

2. The images which private and museum collections make publicly available (e.g., Hatanaka 1996) are sometimes not easy to differentiate from those in auction catalogues (e.g., Tuli 1999), which often attempt to provide stylistic overviews and formulate categories while fetishizing the textile completely as a decontextualized object.

3. The post-Independence Zamindari Bill (concerning land redistribution) abolished the Privy Purse, a system of payments to former royalty, and led to the relative impoverishment of the minor aristocracy, who began to sell off luxury clothing as collectible objects. Shobha Deepak Singh claimed that clothing was the first of the possessions to go, and jewelry the last. Previously there had been little market for antique clothing, because Indians had been more interested in new textiles. During the 1950s and 1960s, many rich textiles entered the antique market and formed the nucleus of Indian museum collections as well as augmenting foreign ones.

4. Lemire refers to crockery sellers in mid-nineteenth-century London who carried baskets of crockery for miles, calling their wares to sell or to barter for used clothing. The trade was recorded as early as 1742 in an Old Bailey trial, and was "a process as old as the story of Aladdin's lamp" (Lemire 1988, 8).

5. Tarlo describes the Waghris as "a low-status group who are scattered throughout the towns and villages of Gujarat. . . . In Gujarat, they have a widespread reputation for being dirty and disreputable, and their name is frequently used as an insult. This reputation probably stems partly from their poverty and lowly social position, and partly also from the fact that from 1879–1952 they were classified by the British as a 'criminal caste,' a label which certainly cannot have boosted their image" (Tarlo 1996b, n. 22).

6. An everyday sari costs Rs 200 to 500, though some could have been considerably more expensive; each suit costs perhaps Rs 500 to 800; and genuine branded jeans cost at least Rs 1,000.

7. In the Indian context, Das has analyzed the use of body symbolism in defining impurity, noting that during mourning, for example, it is the peripheries of the body that are emphasized. Exuviae such as hair and nails both belong to the body and yet are outside it—they are allowed to grow freely, and hair hangs loose, just as clothes are left unknotted, shoes not worn, rings and bangles discarded.

"Play on hair, nails and extremities allows the use of body symbolism to express both the normal containment of categories and a state of liminality. Impurity is a metaphor for expressing liminality," and liminality can symbolize the creative transcendence of the given categories of a system (Das 1990, 127, 131).

8. According to Valeri (1985), Loisy's theory of sacrifice posits its "victims" not as gifts to spiritual recipients who stand in a relationship of unequal reciprocity with the givers (Hubert and Mauss 1964, Mauss 1954), but as icons which mimetically capture what the sacrificer is hoping to achieve. The efficacy of the rite lies in its physicality and immediacy, and its ability to restore a proper relationship between the self at one point in time (the visible realm) and future social relations (symbolized by the realm of the invisible) (see Carter 2003).

9. Kaviraj demonstrates that common space is only defined in terms relative to the home, a conceptual distinction recognized in the phrase *ghar/baahar* ("home/ outside"; Kaviraj 1997), which refutes Habermas's notion of the public sphere (Habermas 1989). Kaviraj notes the British colonial authorities' puzzlement at the apparent contradiction between the spotless cleanliness of upper-caste households (cleaning being a quasi-religious duty) and their sometimes filthy environs. Explaining the difference between the concepts of cleanliness and purity and that of hygiene, he says, "When the garbage is dumped, it is not placed at a point where it cannot casually affect the realm of the household and its hygienic well-being. It is thrown over a conceptual boundary [in relation to the home]. . . . There is a sense that the 'outside' is not amenable to control. . . . The exterior is abandoned to an intrinsic disorderliness" (Kaviraj 1997, 98–99).

10. Brass and copper cooking pots also need to be periodically recoated with tin (*kalai*) to stop food spoiling, another of the reasons for their being replaced by stainless steel.

11. Not only are pots an important part of a dowry, but dowry items are often stored in pot-shaped boxes and containers, often sealed with a lid (Miller 1985, 69). In the Gujarati culture discussed by Jain and Patel (1980), dowries include three main types of metal utensils: *goli* (large pitchers), *katodan* (circular boxes with domed lids), and *karandiya* (boxes with cone-shaped lids). Although, as in most of India, various containers, such as large wooden chests and now tin trunks, are used for storing household goods, the *katodan* is mainly used by Kathis and Rajputs for storing clothes. Since these pots are metal containers, they are analogous to the metal wardrobe. Both encapsulate the clothes within, protecting and valorizing them. Metal is an impervious material, and therefore easier to keep both clean and ritually pure; their use as containers for cloth acts as a barrier to the transference of undesirable substances in either direction.

12. Gell refers to the importance of brassware in the form of pots and plates amongst the Muria Gonds in Madhya Pradesh (Gell 1986, 128–29). His informant Tiri's mother seemed to be unusually prone to collecting them, and enjoyed her public display of such prestigious goods. Their use had been adopted from the local Hindus of higher caste, who themselves were already coming to prefer stainless steel utensils. Gell suggests that Tiri's mother projected her role as Muria matriarch onto her collection, and her favorite pot in particular.

13. Pots are ritually used to represent, contain, and embody divinity (Good 1983, 234), while Hindu beliefs concerning death and cremation rituals also show

the relationship between pots and skulls (Parry 1994). Kramrisch discusses the importance of the ritual pot, the *kalash,* as the quintessentially subsuming form (Kramrisch 1946), and Babb's informant claimed that "it represents the 'finite in the infinite,' that is, a tangible form for the deity during a ritual" (Babb 1975, 42). As containers of spirit and representations of Ram, the ideal husband, pots are worshipped at Karva Chauth (Freed and Freed 1998). They can also be used to sacralize space, as the tower of pots, *cauri,* mark the four corners of the mandala in which wedding ceremonies take place. Miller found that metal pots had assumed some of the ritual significance of clay pots, yet warns against overgeneralization. He argues that pots "frame" action, and are part of the "spread of paraphernalia" which build up the auspicious context of the rites (Miller 1985, 140).

6. ADDING VALUE

1. Earlier versions of sections of this chapter were published in Norris 2004a and 2008.

2. The specifically European clothes that Rohini and I tried to trade were unusual and rejected outright as totally unsuitable.

3. This spread of an "oriental" aesthetic recalls Waghorne's description of Mrs. Bossom's Victorian living hall in Boston, decorated with artefacts from the East (1994, 249–62). Waghorne describes how these "religious things" are incorporated into the home and constitute an ornament of ontology. Such a spread also fits within Breckenridge's paradigm of an imagined *ecumene* (Breckenridge 1989); in her example the nineteenth-century World's Fairs were technologies which created a Victorian *ecumene* that encompassed Great Britain, the U.S., and India.

4. One trader suggested that tourists might feel that stains lent the garment a feeling of age, perhaps that the material was in some way "antique"; at this time nobody was marketing such clothes as recycled products.

5. In contrast, the predicament of being overprepared is the starting point for Anne Tyler's humorously perceptive novel *The Accidental Tourist* (Tyler 1985). Compare this predicament to images of Victorian and Edwardian travelers fully equipped to maintain the status quo wherever they went, or the later development of wash-and-wear clothing for Western travelers, enabling them to look good in every situation (Braun-Ronsdorf 1962–63).

6. Similarly, the "art/artifact" debate has largely left out the materiality of the objects themselves and the phenomenological experience of the people who buy and use them (Graeburn 1976; Clifford 1988; Marcus and Myers 1995; Steiner 1994).

7. The use of irony and play is minimal within the somewhat narrow confines of fashionable Delhi society. Antiques, crafts, and curios are sold in discreet but well-known shops such as the renowned textile business Bharany's, and by more general dealers in markets such as Sunder Nagar, which include sellers of old textiles from royal courts and the upper classes. At the very upper end of the market a very small number of educated, well-traveled, self-reflective designers and boutique owners have begun to adopt Western ideas of kitsch and the social acceptability of the trajectory of commonplace goods from trash to treasure. "Lifestyle" shops in urban villages such as Hauz Khas and Mehrauli are beginning to catch on.

8. Shobha Deepak Singh saw her exhibition as breaking new ground in presenting clothing as a suitable store of value to a wider public.

9. Sonam had had a successful show at Singapore Fashion Week, and had expounded on the pan-Asian elements of his style in an interview with Ravni Thakur for *Images: Business of Fashion* (Thakur 2003).

7. VALUE AND POTENTIAL

1. See Gell's discussion of Lucretius's films or "rinds," which are shed from the surface of things as a result of an internal jostling within the object (Gell 1998, 44–45). Gell links these "rinds" back to the Maori *hau* of the forest, the principle of growth from within that was analyzed by Mauss (1954).

2. Latour also identifies the processes of creating hybrids of nature and culture as forms of translation, networks of Ariadne's thread interweaving ideas of system, structure, and complexity, ignored by the "modern critical stance" that separates the human from the non-human world (Latour 1993).

3. Oxfam's textile sorting, recycling, and exporting facility, Wastesaver, has contacted Sital for ideas about ways to utilize donated saris.

 BIBLIOGRAPHY

Abram, D., D. Sen, H. Sharkey, and G. J. Williams. 1996. *The Rough Guide to India.* 2nd ed. London: Rough Guides.

Adams, K., and S. Dickey. 2000. Negotiating homes, hegemonies, identities, and politics. Introduction to *Home and Hegemony: Domestic Service and Identity Politics in South and Southeast Asia,* 1–30. Ann Arbor: University of Michigan Press.

Appadurai, A. 1981. Gastro-politics in Hindu South Asia. *American Ethnologist* 8, no. 3:494–511.

———. 1986a. Introduction: Commodities and the politics of value. In *The Social Life of Things: Commodities in Cultural Perspective,* 3–63. Cambridge: Cambridge University Press.

———, ed. 1986b. *The Social Life of Things: Commodities in Cultural Perspective.* Cambridge: Cambridge University Press.

———. 1988. How to make a national cuisine. *Comparative Studies in Society and History* 30:3–24.

Attfield, J. 2000. *Wild Things: The Material Culture of Everyday Life.* Oxford: Berg.

Babb, L. 1975. *The Divine Hierarchy: Popular Hinduism in Central India.* New York: Columbia University Press.

———. 1981. Glancing: Visual interaction in Hinduism. *Journal of Anthropological Research* 37, no. 4:387–401.

Baden, S., and C. Barber. 2005. *The Impact of the Second-Hand Clothing Trade on Developing Countries.* Oxford: Oxfam.

Banerjee, M., and D. Miller. 2003. *The Sari.* Oxford: Berg.

Banim, M., and A. Guy, eds. 2001. *Through the Wardrobe: Women's Relationships with Their Clothes.* Oxford: Berg.

Barnes, R. 1997. *Indian Block-Printed Textiles in Egypt: The Newberry Collection in the Ashmolean Museum.* Oxford: Oxford University Press.

Barnes, R., S. Cohen, and R. Crill. 2002. *Trade, Temple and Court: Indian Textiles from the Tapi Collection.* Mumbai: India Book House.

Barthes, R. 1985. *The Fashion System.* London: Cape.

Bataille, Georges. 1985. *Visions of Excess: Selected Writings, 1927–1939.* Ed. and with an introduction by Allan Stoekl. Manchester: Manchester University Press.

Baudrillard, J. 1994. The system of collecting. In *The Cultures of Collecting,* ed. J. Elsner and R. Cardinal, 7–24. London: Reaktion Books.

Bayly, C. A. 1986. The origins of *swadeshi* (home industry): Cloth and Indian society, 1700–1930. In *The Social Life of Things: Commodities in Cultural*

Perspective, ed. A. Appadurai, 285–321. Cambridge: Cambridge University Press.

Bean, S. S. 1989. Gandhi and *khadi,* the fabric of Indian independence. In *Cloth and Human Experience,* ed. A. B. Weiner and J. Schneider, 355–76. Smithsonian Series in Ethnographic Enquiry. Washington, D.C.: Smithsonian Institution Press.

Belk, R. W. 1995. *Collecting in a Consumer Society.* London: Routledge.

Benjamin, W. 1992a. The task of the translator. In *Illuminations,* trans. H. Arendt, 70–82. London: Fontana Press.

———. 1992b. Unpacking my library. In *Illuminations,* trans. H. Arendt, 61–69. London: Fontana Press.

Béteille, A. 1997. Caste in contemporary India. In *Caste Today,* ed. C. J. Fuller, 150–79. New Delhi: Oxford India Paperbacks.

Bourdieu, P. 1977. *Outline of a Theory of Practice.* Cambridge: Cambridge University Press.

Braun-Ronsdorf, M. 1962–63. Travel clothes. *CIBA Review.* Basle, Switzerland: Ciba.

Breckenridge, C. A. 1989. The aesthetics and politics of colonial collecting: India at world fairs. *Comparative Studies in Society and History* 31, no. 2:195–216.

Breman, J. 2003. Labour in the informal sector of the economy. In *The Oxford India Companion to Sociology and Social Anthropology,* ed. V. Das, 1287–1318. Delhi: Oxford University Press.

Campbell, C. 1992. The desire for the new: Its nature and social location as presented in theories of fashion and modern consumerism. In *Consuming Technologies: Media and Information in Domestic Spaces,* ed. R. Silverstone and E. Hirsch, 48–64. London: Routledge.

Carrier, J. G. 1995. *Gifts and Commodities: Exchange and Western Capitalism since 1700.* London: Routledge.

Carter, J., ed. 2003. *Understanding Religious Sacrifice: A Reader.* London: Continuum.

Cerny, C., and S. Seriff, eds. 1996. *Recycled Re-seen: Folk Art from the Global Scrap Heap.* New York: Harry Abrahams in association with the Museum of International Folk Art, Santa Fe.

Chakrabarty, D. 1991. Open space/public space: Garbage, modernity and India. *South Asia* 14, no. 1:15–31.

Chakravarty, R., and N. Gooptu. 2000. Imagi-nation: The media, nation and politics in contemporary India. In *Cultural Encounters: Representing "Otherness,"* ed. E. Hallam and B. Street, 89–107. London: Routledge.

Chaturvedi, B. 1994. *A Tale of Trash: A Survey of Materials, People and Economics Involved in the Recycling Trade in Delhi.* New Delhi: World Wide Fund for Nature India.

Clarke, A. 2000. Mother swapping: The trafficking of nearly new children's wear. In *Commercial Cultures: Economies, Practices, Spaces,* ed. P. Jackson, M. Lowe, D. Miller, and F. Mort, 85–100. Oxford: Berg.

Clifford, J. 1988. *The Predicament of Culture.* Twentieth-Century Ethnography, Literature and Art. Cambridge, Mass.: Harvard University Press.

Cohn, B. 1989. Cloth, clothes and colonialism: India in the nineteenth century. In *Cloth and Human Experience,* ed. A. B. Weiner and J. Schneider,

303–53. Smithsonian Series in Ethnographic Enquiry. Washington, D.C.: Smithsonian Institution Press.

Coomaraswamy, A. K. 1913. *The Arts and Crafts of India and Ceylon*. London: T. N. Foulis.

Coote, J., C. Morton, and J. Nicholson, eds. 2000. *Transformations: The Art of Recycling*. Oxford: Pitt Rivers Museum, University of Oxford.

Corrigan, P. 1989. Gender and the gift: The case of the family clothing economy. *Sociology* 23, no. 4:513–34.

Crooke, W. B. A. 1896. *The Popular Religion and Folk-Lore of Northern India*. Westminster: Archibald Constable.

Cwerner, S. B. 2001. Clothes at rest: Elements for a sociology of the wardrobe. *Fashion Theory: The Journal of Dress, Body and Culture* 5, no. 1:79–92.

Das, P., and S. Bhargar. 1987. Recycling of textiles waste in Ahmedabad. M.A. thesis, National Institute of Design, Ahmedabad.

Das, V. 1976. The uses of liminality: Society and cosmos in Hinduism. *Contributions to Indian Sociology* (n.s.) 10, no. 2:245–63.

———. 1990. *Structure and Cognition: Aspects of Hindu Caste and Ritual*. New Delhi: Oxford University Press.

DEFRA (Department for Environment, Food and Rural Affairs). United Kingdom. 2007. *Waste Strategy for England, 2007*. May 24. London: DEFRA.

Desai, A. V. 1999. *The Price of Onions*. New Delhi: Penguin Books India.

Douglas, M. 1966. *Purity and Danger: An Analysis of the Concepts of Pollution and Taboo*. London: Routledge and Kegan Paul.

Dubey, S. 1992. The middle class. In *India Briefing*, ed. P. Oldenburg, 137–64. Boulder, Colo.: Westview Press.

Dutt, G. S. 1939. The art of the *kantha*. *The Modern Review* 65, no. 4:456–61.

Finch, K. 1996. Textiles as historic documents and their conservation. In *Recent Trends in Conservation of Art Heritage*, ed. S. Dhawan, 189–214. Delhi: Agam Kala Prakashan.

Fisher, T., T. Cooper, S. Woodward, A. Hiller, and H. Goworek. 2008. *Public Understanding of Sustainable Clothing: A Report to the Department for Environment, Food and Rural Affairs*. November. London: DEFRA.

Fonseca, R. 1976. The walled city of Old Delhi. In *Shelter and Society*, ed. P. Oliver, 103–15. London: Barrie and Jenkins.

Freed, S. A., and R. S. Freed. 1998. *Hindu Festivals in a North Indian Village*. Anthropological Papers. New York: American Museum of Natural History.

Geertz, C. 1973. Deep play: Notes on the Balinese cockfight. In *The Interpretation of Cultures: Selected Essays*, 412–53. New York: Basic Books.

Gell, A. 1986. Newcomers to the world of goods: Consumption among the Muria Gonds. In *The Social Life of Things: Commodities in Cultural Perspective*, ed. A. Appadurai, 110–38. Cambridge: Cambridge University Press.

———. 1992a. Inter-tribal commodity barter and reproductive gift exchange in Old Melanesia. In *Barter, Exchange and Value: An Anthropological Approach*, ed. C. Humphrey and S. Hugh-Jones, 142–68. Cambridge: Cambridge University Press.

———. 1992b. The technology of enchantment and the enchantment of technology. In *Anthropology, Art and Aesthetics*, ed. J. Coote and A. A. Shelton, 40–63. Oxford: Clarendon Press.

———. 1998. *Art and Agency: An Anthropological Theory*. Oxford: Clarendon Press.

Gibson, J. J. 1979. *The Ecological Approach to Visual Perception*. Boston, Mass.: Houghton Mifflin.

Ginsberg, M. 1980. Rags to riches: The second-hand clothes trade, 1700–1978. *Costume* 14:121–35.

Goffman, E. 1971. *The Presentation of Self in Everyday Life*. London: Penguin.

Good, A. 1983. A symbolic type and its transformations: The case of South Indian *ponkal*. *Contributions to Indian Sociology* 17:223–44.

Goody, J., and S. Tambiah. 1973. *Bridewealth and Dowry*. Cambridge: Cambridge University Press.

Goonj. 2005. Tsunami: Surplus clothing issue. http://www.goonj.info/Tsunami%20Surplus.php (accessed November 15, 2006).

———. n.d. Vastradaan: A nationwide movement. http://www.goonj.info/vastradaan.php (accessed November 15, 2006).

Gordon, S. 1996. Robes of honour: A "transactional" kingly ceremony. *Indian Economic and Social History Review* 33, no. 3:225–42.

Graeburn, N. H. H., ed. 1976. *Ethnic and Tourist Arts*. Berkeley: University of California Press.

Graedel, T. E., and B. R. Allenby. 1995. *Industrial Ecology*. Englewood Cliffs, N.J.: Prentice Hall.

Gregory, C. A. 1982. *Gifts and Commodities*. London: Academic Press.

Gregson, N., K. Brooks, and L. Crewe. 2000. Narratives of consumption and the body in the space of the charity/shop. In *Commercial Cultures: Economies, Practices, Spaces*, ed. P. Jackson, M. Lowe, D. Miller, and F. Mort, 101–22. Oxford: Berg.

Gregson, N., and L. Crewe. 1997. Performance and possession: Rethinking the act of purchase in the car boot sale. *Journal of Material Culture* 2, no. 2:241–63.

———. 2003. *Second-Hand Cultures*. New York: Berg.

Gregson, N. A. Metcalfe, and L. Crewe. 2007. Identity, mobility and the throwaway society. *Environment and Planning D: Society and Space* 25, no. 4:682–700.

Gudeman, S., and A. Rivera. 1990. *Conversations in Colombia*. Cambridge: Cambridge University Press.

Gupta, M. 2009. Transforming waste into value: Examining resource-recovery practices in Mumbai. M.Sc. dissertation, Anthropology, Environment and Development, University College London.

Guy, J. 1998. *Woven Cargoes: Indian Textiles in the East*. London: Thames and Hudson.

Habermas, Jürgen. 1989. *Structural Transformations of the Public Sphere*. Cambridge, Mass.: MIT Press.

Haggblade, S. 1990. The flip side of fashion: Used clothing exports to the Third World. *Journal of Development Studies* 26, no. 3:505–21.

Hansen, K. T. 1994. Dealing with used clothing: *Salaula* and the construction of identity in Zambia. *Public Culture* 6:503–23.

———. 2000. *Salaula: The World of Secondhand Clothing and Zambia*. Chicago: University of Chicago Press.

——. 2004. Controversies about the international second-hand clothing trade. *Anthropology Today* 20, no. 4:3–9.

Harrison, S. 1995. Anthropological perspectives on the management of knowledge. *Anthropology Today* 11, no. 5:10–14.

——. 1999. Identity as a scarce resource. *Social Anthropology* 7, no. 3:239–51.

Hatanaka, K. 1996. *The Textile Arts of India.* San Francisco: Chronicle Books.

Hawley, J. M. 2006. Digging for diamonds: A conceptual framework for understanding reclaimed textile products. *Clothing and Textiles Research Journal* 24, no. 3:262–75.

Hermann, G. 1997. Gift or commodity: What changes hands in the U.S. garage sale? *American Ethnologist* 24, no. 4:910–30.

Hetherington, K. 2004. Secondhandedness: Consumption, disposal and absent presence. *Environment and Planning D: Society and Space* 22, no. 1:157–73.

Hollander, A. 1993. *Seeing through Clothes.* Berkeley: University of California Press.

Hoskins, J. 1998. *Biographical Objects: How Things Tell the Stories of People's Lives.* New York: Routledge.

Hubert, H., and M. Mauss. 1964. *Sacrifice: Its Nature and Function.* Chicago: University of Chicago Press.

Humphrey, C., and S. Hugh-Jones. 1992. Introduction to *Barter, Exchange and Value: An Anthropological Approach,* 1–20. Cambridge: Cambridge University Press.

Ingold, T. 2000. *The Perception of the Environment: Essays in Livelihood, Dwelling and Skill.* London: Routledge.

Irwin, J. 1973. *The Kashmir Shawl.* London: H.M.S.O.

Irwin, J., and K. B. Brett. 1970. *The Origins of Chintz.* London: H.M.S.O.

Jain, J., and S. C. Patel. 1980. *Utensils.* Ahmedabad: Vishalla Environmental Centre for Heritage of Art, Architecture, and Research and the Gujarat State Handicrafts Development Corporation.

Kaviraj, S. 1997. Filth and the public sphere: Concepts and practices about space in Calcutta. *Public Culture* 10, no. 1:83–113.

Keane, W. 2005. Signs are not the garb of meaning: The social analysis of material things. In *Materiality,* ed. D. Miller, 182–205. Durham, N.C.: Duke University Press.

Kingston, S. 1999. The essential attitude: Authenticity in primitive art, ethnographic performances and museums. *Journal of Material Culture* 4, no. 3:338–51.

Kopytoff, I. 1986. The cultural biography of things: Commoditization as process. In *The Social Life of Things,* ed. A. Appadurai, 64–91. Cambridge: Cambridge University Press.

Korom, F. 1996. Recycling in India: Status and economic realities. In *Recycled Re-seen: Folk Art from the Global Scrap Heap,* ed. C. Cerny and S. Seriff, 118–29. New York: Harry Abrahams in association with the Museum of International Folk Art, Santa Fe.

——. 1998. On the ethics and aesthetics of recycling in India. In *Purifying the Earthly Body of God: Religion and Ecology in Hindu India,* ed. L. Nelson, 197–224. New Delhi: SUNY Press.

Kramrisch, S. 1946. *The Hindu Temple.* Calcutta: University of Calcutta Press.

———. 1989. *Kantha* textiles. In *Handwoven Fabrics of India,* ed. J. Dhamija and J. Jain, 78–83. Ahmedabad: Mapin Publishing.

Küchler, S. 1992. Making skins: *Malanggan* and the idiom of kinship in northern New Ireland. In *Anthropology, Art and Aesthetics,* ed. J. Coote and A. A. Shelton, 94–112. Oxford: Clarendon Press.

———. 1997. Sacrificial economy and its objects: Rethinking colonial collecting in Oceania. *Journal of Material Culture* 2, no. 1:39–60.

———. 2002. *Malanggan: Art, Memory and Sacrifice.* Oxford: Berg.

Küchler, S., and G. Were. 2005. Introduction to *The Art of Clothing: A Pacific Experience,* xix–xxx. London: UCL Press.

Kumar, R. 1999. *Costumes and Textiles of Royal India.* London: Christie's Books.

Lamb, S. 2000. *White Saris and Sweet Mangoes: Aging, Gender and Body in North India.* Berkeley: University of California Press.

Latour, B. 1993. *We Have Never Been Modern.* London: Prentice Hall.

Lemire, B. 1988. Consumerism in pre-industrial and early industrial England: The trade in second-hand clothes. *Journal of British Studies* 27, no. 1:1–24.

———. 1991. *Fashion's Favourite: The Cotton Trade and the Consumer in Britain, 1660–1800.* Pasold Studies in Textile History 9. Oxford: Oxford University Press.

———. 1997. *Dress, Culture and Commerce: The English Clothing Trade before the Factory, 1660–1800.* London: Macmillan.

Lisney, R., K. Riley, and J. Banks. 2003–2004. From Waste to Resource Management. Parts 1 and 2. *Management Services* 47, no. 12:8–14; 48, no. 1:6–12.

Loisy, A. 1920. *Essai historique sur le sacrifice.* Paris: E. Nourry.

Lurie, A. 1981. *The Language of Clothes.* London: Bloomsbury.

Lury, C. 1997. The objects of travel. In *Touring Cultures: Transformations of Travel and Theory,* ed. C. Rojek and J. Urry, 75–95. London: Routledge.

Marcus, G., and F. Myers. 1995. *The Traffic in Art and Culture: Refiguring Art and Anthropology.* Berkeley: University of California Press.

Marriott, M. 1976. Hindu transactions: Diversity without dualism. In *Transactions and Meaning: Directions in the Anthropology of Exchange and Symbolic Behaviour,* ed. B. Kapferer, 109–42. Philadelphia: Institute for the Study of Human Issues.

Marriott, M., and R. Inden. 1977. Toward an ethnosociology of South Asian caste systems. In *The New Wind: Changing Identities in South Asia,* ed. K. David, 277–38. The Hague: Mouton.

Martin, P. 1994. Contesting clothes in colonial Brazzaville. *Journal of African History* 35, no. 3:401–26.

Maskiell, M., and A. Mayor. 2001. Killer *khilats,* part 1: Legends of poisoned "robes of honour" in India. *Folklore* 112:23–45.

Mauss, M. 1954. *The Gift.* London: Routledge.

McDonough, W., and M. Braungart. 2002. *Cradle to Cradle: Remaking the Way We Make Things.* New York: North Point Press.

McRobbie, A., ed. 1989. *Zoot Suits and Second-Hand Dresses: An Anthology of Fashion and Music.* London: Macmillan.

Mehta, R., and R. W. Belk. 1991. Artifacts, identity and transition: Favorite possessions of Indians and Indian immigrants to the United States. *Journal of Consumer Research* 17, no. 4:398–411.

Miller, D. 1985. *Artefacts as Categories: A Study of Ceramic Variability in Central India.* Cambridge: Cambridge University Press.

———. 1987. *Material Culture and Mass Consumption.* Oxford: Blackwell.

———. 1991. Primitive art and the necessity of primitivism to art. In *The Myth of Primitivism: Perspectives on Art,* ed. S. Hiller, 50–71. London: Routledge.

———. 1998. *A Theory of Shopping.* London: Routledge.

———. 2000. Introduction: The birth of value. In *Commercial Cultures: Economies, Practices, Spaces,* ed. P. Jackson, M. Lowe, D. Miller, and F. Mort, 77–84. Oxford: Berg.

Mines, M. 1988. Conceptualizing the person: Hierarchical society and individual autonomy in India. *American Anthropologist* (n.s.) 90, no. 3:568–79.

Mookerjee, P. 1987. *Pathway Icons.* London: Thames and Hudson.

Munn, N. D. 1983. Gawan Kula: Spatiotemporal control and the symbolism of influence. In *The Kula: New Perspectives on Massim Exchange,* ed. J. W. Leach and E. Leach, 277–308. Cambridge: Cambridge University Press.

———. 1986. *The Fame of Gawa: A Symbolic Study of Value Transformation in a Massim (Papua New Guinea) Society.* Durham, N.C.: Duke University Press.

Murray, Robin. 2002. *Zero Waste.* London: Greenpeace Environmental Trust.

Norris, L. 2004a. Creative entrepreneurs: The recycling of second-hand Indian clothing. In *Old Clothes, New Looks,* ed. A. Palmer and H. Clark, 119–35. Oxford: Berg.

———. 2004b. Shedding skins: The materiality of divestment in India. *Journal of Material Culture* 9, no. 1:59–71.

———. 2005a. Cast(e)-off clothing: A response to K. Tranberg Hansen (AT 20[4]). *Anthropology Today* 21, no. 3:24.

———. 2005b. Cloth that lies: The secrets of recycling in India. In *Clothing as Material Culture,* ed. S. Küchler and D. Miller, 83–106. Oxford: Berg.

———. 2006. What makes a textile modern? The recycling of clothing in the Punjabi shoddy trade. In *The Future of the Twentieth Century: Collecting, Interpreting and Conserving Modern Materials,* ed. C. Rogerson and P. Garside, 24–29. Papers presented at the second annual conference of the AHRC Research Centre for Textile Conservation and Textile Studies, July 26–28, 2005. Winchester, UK: Archetype Publications.

———. 2008. Recycling and reincarnation: The journeys of Indian saris. *Mobilities* 3, no. 3:415–36.

Oakdene Hollins Ltd., Salvation Army Trading Company Ltd., and Nonwovens Innovation and Research Institute Ltd. 2006. *Recycling of Low Grade Clothing Waste.* Commissioned by Department for Environment, Food and Rural Affairs (DEFRA), UK. Available at http://www.oakdenehollins.co.uk/pdf/Recycle-Low-Grade-Clothing.pdf (accessed September 16, 2009).

Parry, J. 1982. Sacrificial death and the neophrageous ascetic. In *Death and the Regeneration of Life,* ed. M. Bloch and J. Parry, 74–110. Cambridge: Cambridge University Press.

———. 1986. *The gift,* the Indian gift and the "Indian gift." *Man* (n.s.) 21, no. 3:453–73.

———. 1989. On the moral perils of exchange. In *Money and the Morality of Exchange,* ed. J. Parry and M. Bloch, 64–93. Cambridge: Cambridge University Press.

———. 1994. *Death in Banaras.* Cambridge: Cambridge University Press.

Pearce, S. M. 1992. *Museums, Objects and Collections: A Cultural Study.* Washington, D.C.: Smithsonian Institution Press.

———, ed. 1994. *Interpreting Objects and Collections.* London: Routledge.

———. 1995. *On Collecting: An Investigation into Collecting in the European Tradition.* London: Routledge.

Pellizi, F. 1995. Editorial: Remains. *Res* 27:5–21.

Pinney, C. 1993. "To know a man from his face": Photowallahs and the uses of visual anthropology. *Visual Anthropology Review* 9, no. 2:188–225.

———. 1997. *Camera Indica: The Social Life of Indian Photographs.* London: Reaktion Books.

———. 2002. Visual culture. In *The Material Culture Reader,* ed. V. Buchli, 81–103. Oxford: Berg.

Pomian, K. 1990. *Collectors and Curiosities.* Oxford: Polity Press.

Raheja, G. G. 1988. *The Poison in the Gift: Ritual, Prestation, and the Dominant Caste in a North Indian Village.* Chicago: University of Chicago Press.

Resource Alliance and Nand and Jeet Khemka Foundation. 2007. *India NGO Awards 2007: Celebrating Success . . . Rewarding Excellence.* http://www.resource-alliance.org/documents/casebook_2007.pdf (accessed November 11, 2009).

Riley, K., R. Lisney, C. J. Banks, G. Graveson, I. Avery, P. Archer, S. Read, I. Bartle, R. Read, and R. Freeman, R. 2005. *From Waste to Resource Management: An Update for 2005.* Winchester, UK: Hampshire Printing Services. http://eprints.soton.ac.uk/52684/.

Rivoli, P. 2005. *The Travels of a T-shirt in the Global Economy: An Economist Examines the Markets, Power and Politics of World Trade.* Hoboken, N.J.: John Wiley.

Roy, T., ed. 1996. *Cloth and Commerce: Textiles in Colonial India.* New Delhi: Sage Publications India.

Said, E. 1978. *Orientalism.* New York: Random House.

Seriff, S. 1996. Folk art from the global scrap heap: The place of irony in the politics of poverty. In *Recycled Re-seen: Folk Art from the Global Scrap Heap,* ed. C. Cerny and S. Seriff, 8–29. New York: Harry Abrahams in association with the Museum of International Folk Art, Santa Fe.

Sharma, U. 1984. Dowry in north India: Its consequences for women. In *Women and Property—Women as Property,* ed. R. Hirschon, 62–74. Oxford Women's Series. London: Croom Helm.

Simmel, G. 1971. The stranger. In *Georg Simmel: On Individuality and Social Forms,* ed. and trans. D. Levine, 143–49. Chicago: University of Chicago Press.

Srinivas, M. N. 1956. A note on Sanskritization and Westernization. *Far Eastern Quarterly* 15, no. 4:481–96.

Stallybrass, P. 1993. Worn worlds: Clothes, mourning and the life of things. *Yale Review* 81, no. 2:35–50.

———. 1998. Marx's coat. In *Border Fetishisms: Material Objects in Unstable Spaces,* ed. P. Spyer, 183–207. New York: Routledge.

Steedman, C. 1998. What a rag rug means. *Journal of Material Culture* 3, no. 3:259–81.

Steiner, C. B. 1994. Technologies of resistance: Structural alteration of trade cloth in four societies. *Zeitschrift für Ethnologie* 119:75–94.

———. 1995. *African Art in Transit.* Cambridge: Cambridge University Press.

———. 1999. Authenticity, repetition, and the aesthetics of seriality: The work of tourist art in the age of mechanical reproduction. In *Unpacking Culture: Art and Commodity in Colonial and Postcolonial Worlds,* ed. R. B. Phillips and C. B. Steiner, 87–103. Berkeley: University of California Press.

Stewart, S. 1993. *On Longing: Narratives of the Miniature, the Gigantic, the Souvenir, the Collection.* Durham, N.C.: Duke University Press.

Strathern, A. J., and M. Strathern. 1971. *Self-Decoration in Mount Hagen.* London: Duckworth.

Strathern, M. 1988. *The Gender of the Gift.* Berkeley: University of California Press.

———. 1999. *Property, Substance and Effect: Anthropological Essays on Persons and Things.* London: Athlone Press.

Tarlo, E. 1996a. *Clothing Matters: Dress and Identity in India.* London: C. Hurst.

———. 1996b. Fabricating regional value: Embroidery traders from Gujarat. Paper presented at the Fourth European Modern South Asia Conference, Copenhagen, 1996.

———. 1997. The genesis and growth of a business community: A case study of Vaghri street traders in Ahmedabad. In *Webs of Trade: Dynamics of Business Communities in Western India,* ed. P. Cadène and D. Vidal, 53–84. New Delhi: Manohar.

Thakur, R. 2003. Re-presenting Asia: Indian fashion and the world. *Images: Business of Fashion* (Images Multimedia, New Delhi) 4, no 1:51–52.

Thomas, N. 2000. Technologies of conversion: Cloth and Christianity in Polynesia. In *Hybridity and Its Discontents: Politics, Science and Culture,* ed. A. Brah and A. E. Coombes, 198–215. London: Routledge.

Thompson, M. 1979. *Rubbish Theory: The Creation and Destruction of Value.* Oxford: Oxford University Press.

Tuli, N. 1999. *Intuitive-Logic, the Next Step: Auctions of Indian Art.* New Delhi: HEART, the Tuli Foundation for Holistic Education and Art.

Tyler, A. 1985. *The Accidental Tourist.* New York: Berkeley Books.

Valeri, V. 1985. *Kingship and Sacrifice: Ritual and Society in Ancient Hawaii.* Chicago: University of Chicago Press.

Varma, P. 1998. *The Great Indian Middle Class.* New Delhi: Penguin Books India.

Vatuk, S. 1975. Gifts and affines in north India. *Contributions to Indian Sociology* (n.s.) 9, no. 2:155–96.

Veblen, T. 1994. *The Theory of the Leisure Class.* New York: Penguin Books.

Waghorne, J. P. 1994. *The Raja's Magic Clothes: Re-visioning Kingship and Divinity in England's India*. University Park: Pennsylvania State University Press.

Weiner, A. B. 1992. *Inalienable Possessions: The Paradox of Keeping-While-Giving*. Berkeley: University of California Press.

Weiner, A. B., and J. Schneider, eds. 1989. *Cloth and Human Experience*. Smithsonian Series in Ethnographic Enquiry. Washington, D.C.: Smithsonian Institution Press.

Wolf, E. 1956. Aspects of group relations in a complex society: Mexico. *American Anthropologist* 58:1065–78.

Woodward, S. 2007. *Why Women Wear What They Wear*. Oxford: Berg.

 INDEX

Italicized page numbers indicate illustrations.

TRACKING GLOBALIZATION

ILLICIT FLOWS AND CRIMINAL THINGS
States, Borders, and the Other Side of Globalization
Edited by Willem van Schendel and Itty Abraham

GLOBALIZING TOBACCO CONTROL
Anti-Smoking Campaigns in California, France, and Japan
Roddey Reid

GENERATIONS AND GLOBALIZATION
Youth, Age, and Family in the New World Economy
Edited by Jennifer Cole and Deborah Durham

YOUTH AND THE CITY IN THE GLOBAL SOUTH
*Karen Hansen in collaboration with Anne Line Dalsgaard,
Katherine Gough, Ulla Ambrosius Madsen,
Karen Valentin, and Norbert Wildermuth*

MADE IN MEXICO
Zapotec Weavers and the Global Ethnic Art Market
W. Warner Wood

THE AMERICAN WAR IN CONTEMPORARY VIETNAM
Transnational Remembrance and Representation
Christina Schwenkel

STREET DREAMS AND HIP HOP BARBERSHOPS
Global Fantasy in Urban Tanzania
Brad Weiss

LUCY NORRIS is Senior Research Fellow in the Department of Anthropology, University College London, where she is currently working on issues of global textile waste. She was formerly Collections Manager at the Horniman Museum, London, and has taught material culture, South Asian anthropology, and research methods at University College London, the Royal College of Art and Design, and Chelsea College of Art and Design. She is co-author of *Bali, The Imaginary Museum: The Photographs of Walter Spies and Beryl de Zoete.*